BEYOND NUMBERS

A Practical Guide to Teaching Math Biblically

Katherine [Loop] Hannon
Christian Perspective, LLC • Loganville, GA

First Printing: 2005
Second Printing: 2007
Second Edition: 2011
Third Edition: 2017
Fourth Edition: 2021

Published by
Christian Perspective, LLC
Loganville, GA
www.christianperspective.net

ISBN First Edition: 0-9773611-1-X; 978-0-9773611-1-3
ISBN Second Edition: 978-0-9773611-3-7
ISBN Third Edition: 978-0-9773611-5-1
ISBN Fourth Edition: 978-0-9773611-6-8

Cover and Book Design and Layout: Susannah Miller (susannah.miller@gmail.com)
Layout of Fourth Edition Edits and Adaptions in Layout for New Book Size: Katherine Hannon
Cover Photo: istockphoto.com | Xaviarnau
Illustration Artist: Jeff Cortazo

Unless otherwise marked, "Scripture taken from the HOLY BIBLE, NEW INTERNATIONAL VERSION (North American Edition). Copyright © 1973, 1978, 1984 by International Bible Society. Used by permission of Zondervan Publishing House."

When marked NASB, "Scripture taken from the NEW AMERICAN STANDARD BIBLE ®, Copyright © 1960, 1962, 1963,1968, 1971,1972, 1973, 1975, 1977, 1995 by The Lockman Foundation. Used by permission."

About the Cover: We chose a sunflower because math helps us see the design God placed within the seed arrangements of a sunflower (see page 9). Sunflowers also grow up toward the sun, a reminder of our desire to grow spiritually toward the Son and to seek Him in everything—including math.

*Dedicated to my Aunt Marie
And to the many other
Homeschool moms who will read this book.
May you and your children
Be blessed by seeing God in math.*

CONTENTS

Preface to the Fourth Edition   vii

Preface to the First Edition   ix

Acknowledgements   xi

Where Did Math Come From, and Why Does It Work?   1

Math Points Us to God   7
    God's Greatness   7
    God's Wisdom and Care   8

Math's Practicality   11
    Math Helps Us Explore Creation   11
    Math Leads to Useful Inventions   13
    Math Assists Us in Our Day-to-Day Life   15

Math Is Not Neutral!   19

Harm to the Heart   25

Adopting a New Heart Toward Math   29
    Math Methods from the Past   37
    The Joy of Biblical Math   39

Preparing to Teach Math Biblically   45
    Remind Yourself of God's Truths   45
    Examine Your Goals   46

Teaching Math Biblically   49
    General Guidelines   49
    View Math Biblically   52
    Use Math Practically   54
    Putting It All Together   56

Ready, Set, Now What Do I Do?   61
   In Your Curriculum   63
   In Resources   63
   In Your Own Life   63

Curriculums and Resources   67
   What to Look for/How to Modify   67
   Curriculums   70
   Reference Books   79
   Supplemental Books   81

Overcoming the Difficulties   85

Parting Thought   89

Appendix A: Idea Notebook   91
   Common Ways We Use Math   91
   Common Projects Using Multiple Math Concepts 95
   A Few Basic Ways Math Is Used in Science   96

Appendix B: Bibliography   97
   Modern Textbooks   97
   Older Textbooks   100
   Supplemental Math Books   102
   Miscellaneous   103

PREFACE TO THE FOURTH EDITION

It has been a joy over the years to hear from readers how they've used *Beyond Numbers*. In fact, when the book temporarily went out of print, we quickly found a way to bring it back due to the requests we received.

Beyond Numbers was the very first book I wrote on math from a biblical worldview. I finished it more than a decade and a half ago now. Since then, I've gone on to develop full curriculla for both junior high and high school, eCourses, and a more in-depth guide to help in teaching arithmetic concepts. While I've learned a lot more and about God's handiwork in math over the years and continue to be amazed afresh at the intricacies of His creation and how math shouts out His faithfulness, I believe *Beyond Numbers* still summarizes the essense of how math can be viewed and taught from a biblical worldview. I hope this resource will continue to be a blessing for those looking to understand the big picture quickly. And know that there are resources that go into more depth available if you want to dig deeper. (See ChristianPerspective.net for my other resources.)

This edition is only slightly revised—the main differences being my name being updated to my married name (God having blessed me with a wonderful husband who shares my passion for pointing others to God's handiwork in math) and an updated Curriculums and Resources chapter. I also added more to the geometry section on pages 53-54 and made a few other minor changes. And you'll notice the size of the book is slightly different, as we adapted it to print on demand. But the bulk of the content remains the same.

In Christ,

Katherine Hannon
Loganville, GA
August 2021

PREFACE TO THE FIRST EDITION

Math was always a strong subject for me. My father and grandmother are CPAs. My mom, who homeschooled me since first grade, majored in math. Because my entire family loved math, I learned to enjoy it as well. I progressed rapidly, completing Algebra 2 by ninth grade.

Yet, while I did well in math, I did not view math biblically. I recognized God's presence in other subjects, but not in math. Science clearly pointed to the Creator. History vividly demonstrated God's sovereignty over the affairs of man. But math? I did not understand how God had anything to do with math.

As I progressed to the higher levels of math, I began to wonder why I needed to learn some of these math concepts. How would I ever use geometry or advanced algebra in my own life? I learned the advanced math concepts as one would learn a fascinating game, but I saw few practical applications for what I learned.

In my senior year, my view of math changed entirely after I read a book titled *Mathematics: Is God Silent?* by James Nickel. I started the book rather skeptically, but soon could not put it down. All my old ideas about math crumbled. Math was *not* neutral. God *was* in math. I did not have to simply believe what my textbooks told me. Math, even geometry and algebra, had meaning and application in real life.

After I graduated, God prompted me to write this simple, nontechnical guide to biblical math. I have spent the last two years researching and writing this guide. My appreciation for math and my understanding of how it points to God has continued to grow as I have explored the Scriptures, numerous math textbooks, and *Truth in the Transcendent*, a book on the Bible and math by Larry Zimmerman. Giving piano instruction has also taught me a great deal about teaching and contributed to many of the ideas I will share in this guide. Many of the other teaching ideas came from watching my mom learn and grow as she taught my brother and me.

In this guide you will discover what the Bible says about math. You will explore where math originated and why it works. You will see the harm in the way most textbooks teach math. You will walk through all aspects of teaching math biblically, from preparing and choosing a curriculum, to overcoming obstacles. Throughout the process, you will learn to look beyond numbers and to behold God in math.

May the Lord use this guide to encourage and equip you to look to Him in everything, *including* math.

Katherine Loop

ACKNOWLEDGEMENTS

I am amazed and humbled by the number of people who have helped make this guide possible. I would like to a take moment to thank them all for their assistance and to let them know what a blessing and answer to prayer they have been to me.

First, I wish to thank James Nickel for his book *Mathematics: Is God Silent?* This book showed me the error in "neutral" math and taught me to approach math as God's handiwork. For this, as well as for Mr. Nickel's encouragement with this project, I am truly grateful.

Second, I wish to thank my family, who went out of their way to help me. I especially wish to thank my mother, who has tirelessly helped in so many ways, and my grandfather Charles Loop, who has devoted hours to this project and whose honest opinions have helped me focus and simplify my thoughts.

I owe a special thanks to three talented people, without whom this book would still be an idea: Liz Harp, the editor; Jeff Cortazzo, the illustrator; and Adrienne Butler, the graphic designer of the first edition. Words do not adequately express my thanks to you for freely giving so many hours of your time and God-given talents. I would also like to thank Barbara Coyle for her assistance with editing and Susannah Miller for designing the cover for and laying out the second edition, a layout this edition is based on.

It would be impossible for me to mention by name all those who have, in one way or another, helped me with this guide. Many moms gave of their time to review this project prior to its publication, while others graciously made their math textbooks available for review. The librarians at the National Education Library granted me access to and assisted me in viewing the library's rare 1800 math textbook collection. Numerous people all over the country upheld this project in their prayers and encouraged me to keep going. Many others have indirectly influenced this work by impacting my spiritual life over the years. Thank you to everyone for all you have done!

Above all, I want to thank the Lord for opening my eyes to His presence in math. Compiling this guide has been an exciting journey into God's faithfulness. Without Him, neither math nor this little guide would be possible.

CHAPTER 1

Where Did Math Come From, and Why Does It Work?

We are all familiar with math. As toddlers, we learn to count. In first grade, we learn to add and subtract. We eventually discover multiplication, division, fractions, and decimals. As teenagers, we study algebra, geometry, trigonometry, and perhaps calculus. Some of us even go on to study math in college and graduate school. As adults, all of us use math in various activities and ways.

And whenever we think of or use math, we also think of God and how math honors Him. Right? Not really. Most of us have been taught to look at math as a list of numbers and rules. We picture math as a set of number problems like the ones at the beginning of this chapter. Although we may see God's hand in history and science, we do not even look for God in math. We do not see any eternal harm or good in math.

We view math as a neutral subject. Neutral means indifferent or "not engaged on either side; not aligned with a political or ideological grouping."[1] We regard math as a subject "not engaged" and "not aligned" with either biblical or worldly thinking. Believing that math is independent from God, we approach math as a "safe" subject—a subject we can all see the same way, regardless of our religious beliefs. After all, the equation one plus one equals two ("1 + 1 = 2") works the same way for a Christian, Muslim, Buddhist, Hindu, or atheist. Math is a subject of numbers and facts, and most of us

---

[1] *Webster's New Collegiate Dictionary*, 1974 ed., s.v. "Neutral."

think facts are neutral.

But is math really indifferent and neutral? Is such a thing even possible? The Bible warns us that we are in spiritual warfare (Ephesians 6:10-18). It urges us to guard our heart and to test the spirit behind what we are taught (1 John 4:1; Proverbs 4:23). The Bible does not mention any neutral ground. According to the Bible, nothing can really be neutral. Everything will be presented in either a biblical or a worldly fashion.

So how can math be presented biblically? What does God have to do with math? Join me now in looking at an equation we are all familiar with: "1 + 1 = 2." We will examine where this equation came from and why it works. As we examine this equation's origins and ability to work, we will discover some startling truths applicable to every area of math.

Mathematicians throughout history have developed various theories to explain the origin and consistency of addition. Some have speculated that addition exists by sheer chance. Others have claimed man created addition and addition works because man designed it to work.

Most modern textbooks do not even attempt to offer an explanation for addition's existence. Throughout my schooling, not one of my textbooks ever explained where addition originated or why it works. I eventually came to the conclusion that addition, along with all other math facts, is an eternal, self-existent truth.

The Bible gives us a radically different explanation for addition. While we will not find the word "addition" in the Bible, the Bible offers us principles that apply to addition, as well as every other aspect of math. Look at what just two Bible verses reveal.

The verse "For by him all things were created" (Colossians 1:16), tells us where addition originated. It tells us God created all things. The word *all* includes everything, even math. This does not mean God created the symbols 1 and 2. Man developed those symbols. But those symbols represent a real-life principle called addition that is embedded in everything around us—a principle God created.

Let me briefly illustrate. We have often taken an object, added another object to it, and been left with two objects (such as when we took one cookie and one piece of pie and ended up eating two desserts). It does not matter what we attempt to add, objects add predictably. An accountant adds dollars, a baker adds doughnuts, a retailer adds inventory, and an electrician adds wattage. Yet addition works the same way for all of these people. Why? Because things add in a regular, precise manner. We refer to this regularity as the addition principle.

God created this principle that prevails throughout creation. When God created all things, He determined how they would operate. He chose to make different aspects of creation—from dollars to watts—operate according to the same reliable principle we call addition.[2]

Man has developed different ways to express on paper this consistency God placed around us. Throughout history, cultures have used different symbols to represent quantities. For instance, the Romans used Roman numerals (I, V, X, etc.) instead of our current Arabic system (1, 5, 10, etc.). However, man has never *created* anything in math. He has merely *developed* different symbols and systems to represent the orderly way things add. Addition originated with God.

Now that we know where addition originated, let us look at *why* addition works. We will find the answer in the very next verse of Colossians, which reads, "In him all things hold together" (Colossians 1:17). Since the word *all* includes addition, God not only created addition but also continually holds addition together.

How does God hold addition together? As we saw above, addition is a name for the predictable way objects combine. God holds addition together by keeping objects combining consistently.

Can you imagine what would happen if objects did not consistently

---

[2] It is important to note that not everything follows our standard rules of addition. Two water droplets, when added together, form one larger water droplet (1 + 1 = 1). This presents a huge problem for those who look at math as an independent fact, but makes sense when we look at math as a way of describing the principles God created and sustains. God has different, though equally consistent, principles for governing liquids than He does solids. The apparent contradiction in how liquids combine reminds us that, no matter how well we think we have things figured out, God's laws and universe are more complex than we can imagine! [Adapted from Katherine A. Loop, *Revealing Arithmetic: Math Concepts from a Biblical Worldview* (Fairfax, VA: Christian Perspective, 2009), 27.]

combine in the same way? If the equation "1 + 1 = 2" only worked some of the time? Math as we know it would be impossible! It would be pointless to memorize addition facts and solve problems on paper if we could not rely on those facts and problems to work in the physical world. Ponder for a moment what a miracle it is that math works. Whether we are on the moon or in our own backyard, we can rely on one plus one to equal two. Math is consistent because God consistently holds every part of the universe in its place. God's faithfulness in holding this universe together ensures us that objects will add in a predictable manner and that the equation "1 + 1 = 2" will consistently work. [3]

The Bible reveals a lot about addition. We have only looked at two verses and already we know

❖ where addition came from (God created it) and

❖ why addition works (because God faithfully sustains the universe).

The principles we have discovered about the addition equation "1 + 1 = 2" apply to the rest of math. Like addition, all of math is a way of recording and expressing the laws and relationships God created. Math works because God faithfully holds everything in place. Math is not independent from God. It is not neutral. Math's very existence and ability to work is dependent on God's faithfulness in holding everything together!

God's faithfulness not only makes math useful to us, but it also communicates an important message. The Bible tells us that God chose to establish a covenant with the "fixed laws" of heaven and earth (which would include mathematical laws like addition). His faithfulness in keeping His covenant with these laws demonstrates the same faithfulness He will show in His covenant of redemption.

If I have not established my covenant with day and night and the

---

[3] This does not mean that God cannot perform miracles. He can and does! When Jesus fed 5,000 with 5 loaves of bread and 2 fish, those loaves and fish certainly did not add according to their standard fashion. We need to understand that God is consistent to His nature. In His mercy and faithfulness, He governs this universe in a predictable pattern. Out of His same mercy and faithfulness, He sometimes chooses to perform a miracle. Miracles help us see God's sovereignty. Miracles show us God is the One in charge! (Adapted from Loop, *Revealing Arithmetic*, 27.)

fixed laws of heaven and earth, then I will reject the descendants of Jacob and David my servant.

Jeremiah 33:25b, 26

The way God holds everything together according to fixed laws demonstrates that He is a covenant-keeping God. His consistency in keeping one plus one equaling two serves as a continual testimony to us of His loyalty, faithfulness, and loving kindness.

Just think, every time you use math to count or add something, you witness the fact that God is still consistently sustaining the universe. It is as though math were shouting out to you, "God is still in charge! He is faithful. He will keep His covenant of redemption with you."

Math is really a testimony to God's faithfulness and power. Math should continually remind us of God's consistency and trustworthiness. After all, one plus one consistently equals two because an awesome, never-changing God created and sustains all things.

Math is not a neutral subject. As we understand biblically where math came from and why it works so consistently, we realize God is not only in math, but the One who makes math possible. Instead of setting God aside when we approach math, let us set aside our old ideas about math and approach math expecting it to glorify the Lord.

CHAPTER 2
Math Points Us to God

Since God created and sustains math, we should expect to see Him in math. Much as a painting reveals many of its painter's characteristics, math reveals many of God's characteristics. As we saw in the last chapter, math's consistency testifies to God's faithfulness. Here are a few other ways math points to God.

GOD'S GREATNESS

Throughout math, we see an incredible system at work. Just think about how long it takes us to learn even basic math skills. Most of us spend years trying to grasp a small percentage of math's many rules.

As mentioned in the last chapter, math records the orderly way God determined things would interact throughout the universe. God created the complex structure of mathematics and holds it in place. Think about what this means for a moment. God created a universe that interrelates in such a complex and precise manner that mathematicians spend a lifetime trying to understand and record it. God's greatness knows no bounds!

Our finite minds can hardly comprehend some portions of math, such as infinite numbers. Yet God holds infinity in the palm of His hands. The Bible tells us that His thoughts toward us are infinite in number (Psalm 139:17, 18; 40:5). Infinite numbers can only exist because a God of infinite proportions created the universe. As hard as it is for us to comprehend, God is more infinite than infinity![4]

Math's complexities should cause us to turn our eyes to God in awe and wonder, for God created and holds math in place. Math serves as a reminder to us of how much greater God is than we could ever imagine. He is truly beyond our understanding (Job 36:26).

GOD'S WISDOM AND CARE

Math not only shows us God's greatness; it also testifies to God's wisdom and care. Math displays aspects of the universe that our limited senses cannot readily observe. As we use math in conjunction with science to explore the universe, math unveils a wondrously detailed design. This design reminds us that our great God is also a wise and caring God who cares for the tiniest details of His creation.

For example, math exposes something about a honeycomb the naked eye misses. We can observe the bee storing honey in honeycombs made out of hundreds of tiny hexagons. This fact does not appear noteworthy until we bring math into the equation. If we analyze various shapes with differential calculus, we will discover that, of all the shapes in the universe, the hexagon holds the maximum content in the minimum space. It is also sturdy, enabling it to handle the weight of the honey. God designed the bee to store its honey in an extremely efficient manner.[5] What a caring, thoughtful Creator we have!

Math uncovers some amazing design within our own bodies. A computer once did the math necessary to determine what shape would allow a red blood cell to best diffuse oxygen throughout the blood stream. Surprise!

---

[4] God created the infinite galaxies of space. Even the highest heavens cannot contain Him! "But will God really dwell on earth? The heavens, even the highest heaven, cannot contain you. How much less this temple I have built!" (1 Kings 8:27)

[5] James D. Nickel, *Mathematics: Is God Silent?* 2001 rev. ed. (Vallecito, CA: Ross House Books, 2001), 211.

You have thousands of red blood cells flowing through your veins right now in the absolutely ideal shape.[6] By revealing this hidden order, math reminds us that God, not random chance, created us.

Math shows us something remarkable about how we grow. A baby's head is proportionally larger when compared to the rest of his body than an adult's head is. When we grow, our head does not continue to expand at the same rate as our arms and legs—if it did, we would look rather strange and probably be unable to walk. God, in His infinite wisdom and care, designed our bodies to change proportions, giving us exactly what we needed for different stages of life. Babies fall a lot, so babies need extra padding on their heads. As we grow, we need larger feet and muscles to support our growing weight—and that is exactly what God gives us.[7]

Math often exposes order in sounds our ears cannot even hear. By using math to analyze sound, scientists discovered that every time we hear a sound, we actually hear an orderly yet invisible wave. Even more incredibly, these waves all express a pattern orderly enough to be recorded algebraically.[8] Who would have guessed God hid such fascinating order in sound?

Even examining a common sunflower mathematically reveals some surprising news. Did you know that the seeds in a sunflower are arranged in spirals? Some spirals go in one direction, and the others in another. Regardless of how many seeds the sunflower contains, the ratio between the number of spirals in each direction is approximately the same. This ratio enables any given sunflower to hold the maximum number of seeds.[9]

When we look at the heavenly bodies using math, we again see evi-

---

[6] Allen L. Gillen, *Body by Design* (Green Forest, AZ: Master Books, 2001), 76.

[7] Adapted from Loop, *Revealing Arithmetic*, 155.

[8] Nickel, *Mathematics: Is God Silent?* 238-239.

[9] This ratio has been named the Fibonacci sequence and is also present in many other flowers and parts of nature.

dences of God's care. Math helps us see how God hung the heavenly bodies precisely where they need to be—and at precisely the right orbits—to sustain life on earth.

These are just a *few* examples of how math makes the complex design God placed around us visible. When we use math to explore the world, we will never cease to be amazed at the remarkable order it uncovers. Over and over, we will find ourselves awed at our Creator's wisdom and care. As we see God's care over creation, it should encourage our own hearts to trust Him, for "If that is how God clothes the grass of the field, which is here today and tomorrow is thrown into the fire, will he not much more clothe you, O you of little faith?" (Matthew 6:30).

At the same time, examining the complexities around us also reminds us of our need for redemption. This universe, while it still declares the glory of its Creator, is fallen and marred from the perfection God had originally set in place. For instance, while designed to reproduce with the most efficient arrangement of seeds in their spirals, sunflowers in nature have spirals slightly distorted by outside influences—seeds are not all exactly the same size, and external forces such as pressure against other flower buds during seed development may affect the spirals. While math helps us see a general relationship God put into sunflowers that enables them to reproduce quite efficiently, it also reminds us of the universe's fallen condition. This in turn should cause us to rejoice that God has provided us with a way of salvation—Jesus Christ our Lord!

Whether we are studying math or using math to investigate creation, math continually urges us to trust God. Within math, we see a complex system that testifies to God's greatness. When using math to explore the physical world, we discover an intricate design that points us to God's wisdom and care and cause us to rejoice in His salvation.

Every part of math causes us to exclaim with the Psalmist, "Great is the LORD and most worthy of praise; his greatness no one can fathom" (Psalm 145:3).

CHAPTER 3

Math's Practicality

Biblical principles affect more than just the way we see math. They also affect the way we use math. Once we realize that God created and sustains math, we will no longer regard math as meaningless symbols on a piece of paper. We will recognize that those symbols represent principles in the physical world and will consequently no longer be content to use math merely in a textbook. We will want to begin to use math practically in our own lives.

By creating a methodical universe and giving us the mental capacity to record the universe mathematically, God gave us a valuable provision to help us in our lives here on earth. We can think of math as a God-given, multi-purpose tool. Much as a shovel helps us dig deep into the ground and plant gardens or build buildings, math helps us explore the universe to new depths and develop useful inventions. Math also assists us in many daily tasks, in a similar way to an ever-handy screwdriver or pocketknife that assists us in thousands of little activities. In this chapter, we will take a look at a few of the practical ways math helps us.

MATH HELPS US EXPLORE CREATION

Math shows us the design hidden from our senses, such as the design in beehives, blood cells, sound waves, and sunflowers. It also aids in understanding many other aspects of creation.

For example, have you ever wondered how we know about atomic structure? No one has ever seen atoms. They are too small. We know everything we know about them from—you guessed it—math! Scientists used mathematical principles coupled with experimentation to discover the structure of atoms (nucleus, protons, and electrons) and to understand the principles by which atoms operate.

Or have you ever wondered how we know the distance to a star? Obviously, we cannot simply measure that distance with any physical tool of measurement. We use math to estimate a star's distance.

Throughout history, men have used math to describe the consistent way the universe operates. Sir Isaac Newton used math to record the steady attraction objects have toward earth (called the law of gravity). He then applied his law of gravity to find the weight of the sun.[10] Johannes Kepler used math to discover the way planets orbit before modern spacecraft made it possible to actually see the planets in orbit firsthand. The ancient Mayan civilization used math to keep track of the movement of the heavenly bodies.[11] From the beginning of time, math has aided man

---

[10] Irving Adler, *The Giant Golden Book of Mathematics: Exploring the World of Numbers and Space* (NY: Golden Press, 1960), 86.

[11] The Mayans kept detailed almanacs chronicling astronomical movements and basing calendars off these movements. Michael Hoskin, ed., *Cambridge Illustrated History: Astronomy* (Cambridge, UK: Cambridge University Press, 1997), 14-17.

in investigating the earth.

In many different ways, math helps us better comprehend the extraordinary world God created. In the process, it points us to the greatness of the God who created both the microscopic order of the atom and the vast distances of space.

MATH LEADS TO USEFUL INVENTIONS

Nearly all inventions rely on math and the laws (i.e., consistent ways God holds things together) math uncovers. Math is instrumental in designing bridges. Computers are based on a mathematical system of zeros and ones. The strings in a piano and other stringed instruments all mathematically relate to one another. The list of mathematically based inventions goes on and on.

Often, as we use math to help us explore creation, math indirectly leads to inventions. For example, understanding the atom enables us to predict chemical reactions, thus opening the door for developing all types of inventions. Knowledge of the atom has aided in manufacturing synthetic materials such as plastics, explosives such as the atomic bomb, various medical procedures such as x-rays, and even household items such as super glue.[12] Since electricity is based on atoms and how they interact, knowledge of the atom has given us a better understanding of electricity, leading to many of the electrical gadgets we have today. By teaching us about the atom, math has affected every field and aspect of modern life in one way or another.

Likewise, the discovery of the law of gravity has served a critical role in numerous inventions, especially in the aviation field. Because we can determine the pull of gravity, designers are better able to design aircraft, spaceships, and rockets to cope with the force of gravity they will encounter.

The use of mathematics in practical inventions is not limited to our Mod-

---

[12] Super glue is based on molecular bonding. See Neil Schlager, ed., *How Products Are Made: An Illustrated Guide to Product Manufacturing*, vol. 1 (Detroit: Gale Research Inc., 1994), s.v. "Super Glue."

ern Era.[13] Many ancient civilizations, including the Egyptians and Greeks, used various types of sundials (which are based on math) to keep track of time.[14] The Egyptian pyramids demonstrate incredible mathematical usage, as do many of the old European cathedrals and buildings. If you have any doubt of math's practical use historically, just consider the following list of inventions the Greek mathematicians in Alexandria, Egypt produced over two thousand years ago:

Pumps to bring up water from wells and cisterns, pulleys, wedges, tackles, systems of gears, and a mileage-measuring device no different from what may be found in the modern automobile were used commonly. Steam power was employed to drive a vehicle along the city streets in the annual religious parade. Water or air heated by fire in secret vessels of temple altars was used to make statues move. The awe-struck audience observed gods who raised their hands to bless the worshippers, gods shedding tears, and statues pouring out libations. Water power operated a musical organ and made figures on a fountain move automatically while compressed air was used to operate a gun. New mechanical instruments, including an improved sundial, were invented to refine astronomical measurements.[15]

Quite an impressive list. Pumps, gears, mileage-measuring devices, steam powered vehicles, moving statues, musical organs, a compressed air gun—the Alexandrians found numerous practical uses for math!

------- NOTE -------

Alexandrian Greek mathematics differed drastically from the Classical Greek mathematics we will discuss later. While the Classical Greek mathematicians rarely used math practically,

---

[13] For a further listing of different ancient inventions, see Peter James and Nick Thorpe, *Ancient Inventions* (New York: Ballantine Books, 1994).

[14] *The New Encyclopedia Britannica*, 15th ed., s.v. "Sundial." and R. Newton Mayall and Margaret W. Mayall *Sundials: Their Construction and Use*, 3rd ed. (Cambridge, MA: Sky Publishing Corp., 1994).

[15] Morris Kline, *Mathematical Thought from Ancient to Modern Times* (New York: Oxford University Press, 1972), 103, quoted in Nickel, *Mathematics: Is God Silent?* 39.

the Alexandrians, as you can see, used math quite extensively.

Unfortunately, though they used math practically, the Alexandrians enthroned man's reason and missed out on seeing God in math. The Alexandrians serve as a warning to us not just to teach math as a practical tool. The practical element alone does not make math biblical.

MATH ASSISTS US IN OUR DAY-TO-DAY LIFE

In addition to providing us with useful inventions, math assists us in many different situations we encounter throughout our lives. Whatever our occupation, we will find math a useful tool.

Archimedes, a famous Alexandrian Greek mathematician, frequently used math to solve real-life dilemmas. Once, a king wanted to know whether some silver had been added to a gold crown he had ordered. Silver weighs less than gold, so Archimedes knew that if the crown had silver in it, it would be larger (and take up more space) than a solid gold bar of the same weight. By comparing the space the crown took in a bucket of water with the space taken by a solid gold bar of equal weight, Archimedes proved that the crown contained silver. According to legend, Archimedes also used mathematically designed mirrors to reflect and intensify the sun's rays onto enemy warships, causing the ships to catch fire.[16]

Although most of us will not end up solving a golden mystery or setting fire to a fleet of ships, we can still use math to solve problems in our own life. For instance, we can measure the doorway to see if the new couch will fit through it *before* the delivery man arrives. Below are some other common ways you may have used math to help you with everyday dilemmas.

While Shopping

Rounding can be helpful when trying to

---

[16] Nickel, *Mathematics: Is God Silent?* 40.

purchase items within a specific budget. My mother would typically keep a running estimate in her head when shopping to avoid last-minute surprises, especially when paying for items with cash. She unconsciously rounded the price of each item she placed in her cart and kept a mental tally of the total.

Shopping uses much more math than just rounding. I remember being awestruck the first time I shopped for groceries by myself when I discovered how many different brands of noodles the store carried. Which should I buy? I grabbed the cheapest package I could find. My mom later pointed out to me that the least expensive is not necessarily the best deal.

Had I used math to find the price per ounce, I would have realized I paid more per ounce for my noodles than I would have had I chosen a different package.

Keep your eyes open next time you go shopping and see how many different math concepts you use. You may be surprised!

While Cooking

Now you have finished your shopping. You come home to a hungry family. You throw together a batch of cookies and decide to double the recipe so you can bring some cookies to soccer practice. You mentally double ¾ and know you need 1½ cups of flour. Your ½ and ¾ measuring cups are still being washed from lunch, so you use your 1 and your ¼ measuring cups instead, remembering to fill the ¼ cup twice. You have just added and multiplied fractions.

While Repairing or Improving Your Home

Home repairs and improvements require numerous math concepts. Just the other day my mother mixed wallpaper remover with water. If she had not mixed the solution in the correct mathematical proportions, the solution might have been too strong and damaged the wall.

If you have replaced a floor, you have used a lot of math. Finding the area of the floor, deciding how much material to order, comparing prices between stores and different types of flooring, calculating the cost of grout and other supplies, and factoring in the cost of labor all require basic math.

Many other home improvement projects also require math. Determining how many rolls of wallpaper to purchase, what size piece of wood is needed for a project, how many square/linear feet of countertops to replace, or how large a vanity will fit in the bathroom are just a few other home improvements that involve math.

When Buying a Car or House

Math definitely comes in handy when buying a car or house. Math can help you figure out what your monthly loan payment will be. It can also aid in comparing interest rates and financing options.

We all use math frequently in our daily lives. Math is a useful tool for you and your children, not just for engineers or rocket scientists.

Seeing the many ways math can be used to help us explore the earth, invent, and live day-to-day is exciting. Math is *not* a meaningless pile of numbers. It is a real, useful tool!

Yet math would not be useful at all if it did not always work, would it? We could not use math to invent or help us in real life if math only worked some of the time. Math is useful because the universe is consistent, operating according to "fixed laws." And those laws hold true because God consistently holds all things together. Math's usefulness points us to God's faithfulness!

CHAPTER 4
Math Is Not Neutral!

When I first began to recognize the testimony to God within math, I began to wonder why I had not seen God in math before. Why had I gone through school viewing math as nothing more than a list of rules and numbers? How could I have been so blind?

Looking back, I see how my view of math as a neutral set of facts blinded me to God's presence in math. I simply believed whatever my math books taught me. In my mind, math was a neutral subject, a subject that contained neither biblical nor worldly thinking.

I did not realize that, by viewing math as a neutral subject (a subject "not engaged" and "not aligned"[17] with either biblical or worldly thinking), I was really embracing worldly thinking. I was learning math as a subject independent from God. If math is independent from God, then all the incredible order and consistency in math points us, not to the Creator, but to math and human intellect. This view of math is not just unbiblical; it is actually naturalistic and humanistic. It teaches us to trust the creation (math and human intellect) instead of the Creator. My "neutral" view of math was really a very worldly view, steeped in unbiblical thinking.

When you think about it, we cannot really view anything neutrally. The way we see everything will be automatically sifted through a biblical or a worldly philosophy. Even facts are interpreted through a belief system.

For example, consider the media. Newspapers and newscasts purport

---

[17] *Webster's New Collegiate Dictionary*, 1974 ed., s.v. "Neutral."

to report facts. Yet a conservative and a liberal reporter present facts differently. Even when reporting on the same subject, they will often choose to include different facts in their reports. They will unconsciously slant their presentation to their belief about the subject. If you doubt this, just compare a conservative and liberal paper around election time. Somehow, each paper will manage to portray the facts so as to favor the candidate it prefers. Philosophies, or beliefs, come through in the way reporters select and report facts.

For another example, suppose an evolutionist and creationist were both to announce finding a fossil of a previously unknown animal. They would both report this fact in a radically different way. One might describe the fossil as the missing link, while the other might describe it as an animal that has gone extinct since the flood. One report presents the fossil from the worldly philosophy of evolution; the other presents it from the biblical philosophy of a worldwide flood.

In a similar, although more disguised way, underlying beliefs come across in math. Whether we choose to acknowledge it or not, math facts are presented based on an overall belief system. This belief system will either be biblical or unbiblical. There is no other option.

Everything we study either presents God's view or the world's view. We could summarize these two views by the words "dependent" and "independent." God's Word teaches us that we are dependent on God, while all other philosophies claim that, to one extent or another, we can operate or please God on our own, independently from God.

As Christians, we are in continual warfare. We constantly have to fight our tendency to live and think independently from God. The Bible never excludes math from this combat zone. In math, as in everything else, we need to guard against independent thinking—any sort of thinking that encourages us to trust ourselves, operate independently from God, or view something with our own human reason.

Consider the following math lessons.[18] As you read them, ask yourself whether they present a biblical, dependent view of math and life or a worldly, independent one.

Counting is a math skill that we learn early. Counting by ones, we say the numbers 1, 2, 3, 4, 5, 6, ...

These numbers are called **counting numbers**.[19]

Searching for a missing factor is called **division**. A division problem is like a miniature multiplication table. The product is inside the box. The two factors are outside the box. One factor is in front and the other is on top.[20]

The facts stated in the above math lessons are all undoubtedly true. Counting is a math skill we learn early. One, two, and three are called counting numbers. Searching for a missing factor is called division. These lessons certainly appear spiritually neutral.

But look beyond the numbers and facts presented and ask which view

---

[18] Disclaimer: The following example lessons as well as the ones that appear later in the book were chosen from curriculums I had on hand and that I felt accurately represented typical math presentations. It is not my intent to condemn these curriculums over others. It is my intent to show the reader how "neutral" math lessons really present math through a worldly philosophy. This is true for many math curriculums, not just the ones from which the examples were taken.

[19] Stephen Hake and John Saxon, *Math 65: An Incremental Development*, 2nd ed. (Norman, OK: Saxon Publishers, 1995), 1.

[20] Ibid., 74.

of math—a dependent or an independent one—is being taught. While the above lessons present true math facts, they also leave out critical information. We are not told *why* these facts work, nor are we taught how to use them in real life.

Rather than presenting counting and division as principles created by God and as practical tools, these lessons leave the student thinking of counting and division as man-made rules and neutral facts—facts independent from God. They subtly teach the student to view math as an independent, self-existent truth.

This view of math goes directly against Scripture and subtly presents a completely unbiblical philosophy on life. Think about it. If math is independent from God, then it does not require God's help to work. Yet math works with miraculous consistency. We can clearly rely on math. If God is not responsible for math's ability to work, who is? By simply presenting facts without providing an explanation for their existence, this lesson leaves us thinking that either math works all by itself, or else that mathematicians somehow invented math and enable it to work.

Either way, something or someone other than God has received the credit for math's consistency. Math has been used to point us, not to God's faithfulness, but to math's faithfulness and man's ability to reason independently from God.

Do you see the subtle but dangerous philosophy hidden in these math presentations? Teaching math as something independent from God really stems from a philosophy claiming we do not need God—that we can determine truth and explain the universe on our own, using human reason and mathematical principles.

Let us look at one more example of a typical math lesson. Once again we will find that, by ignoring God, this lesson subtly indoctrinates the student in an independent view of math and life in general.

A property describes the way something is. We can't change properties. We are stuck with properties because they are what they are.[21]

---

[21] John H. Saxon, Jr., *Algebra 2:An Incremental Development*, 2nd ed. (Norman, OK: Saxon Pub-

Again, there are some true facts in this math lesson. We cannot change properties. Yet, why can we not change properties? Why do properties work?

"We are stuck" and "they are what they are" clearly give the impression that math properties sustain themselves. They leave us viewing a property as something independent from anything or anyone else.

In contrast, in a biblical perspective, a property is really just one of the "fixed laws" by which God has chosen to sustain the universe. Far from being independent *from* God, a property is entirely dependent *on* God! When we study properties, we ought to be praising the Lord for His faithfulness. Like the other lessons, this apparently "neutral" lesson robbed God of His proper place in math.

Do you see how the above math lessons were not really spiritually neutral? Each one, by presenting math as independent from God, subtly taught a worldly philosophy on life. They taught us to view math through *independent* eyes. These same independent eyes applied to other areas of life teach us that we can trust ourselves, we can determine truth apart from God, and we can save ourselves or please God on our own. The independent thinking in these math lessons, though subtle, is very dangerous.

In contrast, the Bible tells us God created and sustains all things, meaning math is dependent on God. In fact, the Bible's whole message is one of dependence. Over and over, the Bible urges us to stop trusting our works, our wisdom, and ourselves. It continually stresses our need to depend on Christ alone in every area of our life. The Bible even compares us to a branch and tells us that, much like a branch can do nothing without the vine, we can do nothing apart from God.[22] When taught biblically, math should echo this message of dependence and encourage us to depend on our faithful Lord.

<hr>

lishers, 1997), 8.
[22] John 15:4-5

Math cannot really be presented neutrally. It will be presented in one of two ways:

❖ *Dependent* on God, i.e., something He created and sustains; or
❖ *Independent* from God.

Math either points toward God or away from Him. It either encourages us to depend on God or to live independently from God. It either acknowledges God or rejects Him. But math cannot be neutral. If we claim to view math lessons neutrally, we blind our heart to God's presence in math and unconsciously swallow unbiblical, independent thinking.

CHAPTER 5
Harm to the Heart

Since we are so accustomed to viewing math as a neutral subject, I would like to spend a little more time showing the danger of this viewpoint. Let me share with you a few of the ways my belief in math's neutrality harmed me for many years.

Because I believed in math's neutrality, I blindly believed everything my math textbooks told me. I did not look at math with discernment as I did my other subjects. While in other subjects I questioned my textbooks and checked out their teaching against Scripture, I simply accepted everything my math book said as infallible.

I knew I could rely on math to always work. This consistency should have reminded me to trust God; instead, it reminded me to trust math and my math book.

When I entered my teenage years, I began to struggle to make my faith my own. My view of math hampered me in this process. I knew evolutionists claimed they had mathematically proven the earth was millions of years old. If math always worked, should I not believe what these highly learned mathematicians told me?

If I had known the biblical truth about math, I would have seen math as a tool to record God's orderly world, not as an infallible means to determine the earth's origins. I would have known that math's very existence, in fact, testifies to the fact that an awesome Creator, not the random process of evolution, holds this earth together.[23] But because of my belief in

---

[23] For more information on the creation/evolution issue, visit Answers in Genesis' website, www.

25

math's neutrality, my math studies instead subtly encouraged me to trust math, mathematicians, and human intellect.

Math is like a living monument, urging us to stand in awe of whatever keeps it working so marvelously. Math is a truly amazing and complex subject. Numbers all interrelate consistently and predictably. We can use math, year after year, time after time, and it always works. Whether consciously or unconsciously, we give the credit for math somewhere. Math will always encourage us to trust either God or something else.

As we saw in the first two chapters, math should remind us to trust God. We should see math as a record of the incredible consistency with which God sustains all things. We should see God in math. Math should be a reminder of—a monument to—God's faithfulness and wisdom.

When we view math as something neutral or independent from God, math encourages us to trust math itself and our own human wisdom and reasoning. We no longer see math as a monument to God. Instead, we see math as a monument to man and numbers.

There are also other ways a belief in math's neutrality harms our hearts. I remember the confusion I felt as I entered high school. Math seemed like a big mystery to me. Why did math work? Where did math come from? My textbooks never really told me. I found myself memorizing rule after rule without really understanding how that rule came about. I began losing sight of the purpose behind learning math. How would I ever use exponents and algebraic division in my own life?

Had I understood that math was not neutral—that math merely records the order God created—math would not have been a mystery to me. I would have realized that the rules simply record the complex principles by which God holds everything in the physical universe in place. Had my textbooks presented math biblically, they would have taught me exponents and algebraic division as useful, God-given tools. They would have shown me math's meaning and purpose.

Sadly, many students today fail to see math's purpose. They do not realize that math is a way of recording real-life principles. Instead, they wonder, "Why do I have to study math?"

In short, viewing math as a neutral subject causes us to exchange the truth of God for a lie. It causes us to unconsciously adopt independent thinking and rely on man's capabilities and intellect.

Chapter 1 of Romans reminds us that God's creation testifies to Him. We have seen that math is no exception to this. Each and every part of math testifies to God's faithfulness, greatness, and care.

> For since the creation of the world God's invisible qualities—his eternal power and divine nature—have been clearly seen, being understood from what has been made, so that men are without excuse.
>
> Romans 1:20

However, a few verses later God tells us that we would ignore Him and exchange His truth for a lie.

> They exchanged the truth of God for a lie, and worshiped and served created things rather than the Creator—who is forever praised. Amen.
>
> Romans 1:25

Math is not a neutral subject. God's invisible qualities are clearly seen in math. If we are not worshiping God in math, we are unconsciously worshipping the creation, in this case, math itself. We are exchanging God's truth for a lie. Our attempts at neutrality only harm us and open up our hearts to worldly, independent thinking.

CHAPTER 6
Adopting a New Heart Toward Math

If the typical way of teaching math harms the heart, how should we teach math? How can we teach math biblically?

I used to think teaching math biblically meant adding a Bible verse or two to a math lesson. If we start our math lessons with a Bible story or scriptural truth, will we have taught math biblically?

Let us try it. Below is one of the lessons we looked at previously.

> Searching for a missing factor is called **division**. A division problem is like a miniature multiplication table. The product is inside the box. The two factors are outside the box. One factor is in front and the other is on top.[24]

Now we will add the following Scripture verse and thought to the lesson.

> For the word of God is living and active. Sharper than any double-edged sword, it penetrates even to **dividing** soul and spirit. (Hebrews 4:12)

> God's Word is powerful and can **divide** our very soul.

Did we make the lesson biblical? We brought in a truth from God's Word, but we did not change the way division itself is presented. The student is not told that God created and sustains division. Rather, he is left

---

[24] Hake and Saxon, *Math 65*, 74.

viewing division as something independent from God—as a self-existent or man-made fact. This lesson still presents math as something independent from God.

Remember, math will always be taught in one of two ways:

❖ Dependent on God, i.e., something He created and sustains; or
❖ Independent from God.

Although including Bible verses and thoughts might seem like a good idea, these inclusions alone do not teach math biblically. They may encourage the student to depend on God and His Word for his spiritual needs, but they do not teach him to depend on God and His Word in *math*. They do not portray math itself as something dependent on God. Adding a Bible verse or thought to a typical math lesson still teaches math the world's way. It still depicts math as an independent fact. The student is still taught to think about math as a subject independent from God. Dangerous, independent seeds are still being planted.

Below is an example from a Christian textbook of trying to add the Bible to what is still ultimately an independent presentation of math.

Thought: Just as God is our Foundation, so real numbers are the foundation of algebra.[25]

Can you see the harm in this thought? Without intending to, this thought actually reinforces the wrong view of math. It implies that God is our foundation, but real numbers, not God, are the foundation of algebra. The student is taught to look at algebra as a subject independent from God instead of entirely dependent on Him. Far from revealing God in math, this sort of teaching actually encourages the student to trust God in every area *but* math.

Simply adding Bible verses or thoughts to our math lessons does not make math biblical. Math still comes across as independent from God.

Here is an illustration. If you saw a bomb ready to explode, what would you do? Would you tape Bible verses over the bomb until you could no longer see the danger? Of course not! That would be foolish. You would

---

[25] Accelerated Christian Education, *Algebra II*, 1995 rev. (USA: Accelerated Christian Education, 1993), pace 1121, 10.

seek to defuse the bomb.

In math, independent thinking is like a bomb ready to explode. Adding Bible verses to our math lessons will not change the dangerous thinking. The danger can only be "defused" by rejecting the independent thinking and embracing a biblical, dependent view of math.

While merely adding Scripture verses to math encourages students to build their life biblically, it still teaches them to build their thinking about math on a worldly, independent philosophy. To visualize the foolishness of this, picture a house being built. Can you imagine a builder laying a firm foundation on which to build most of the house, but deciding to build a window on the nearby sand? How foolish! Is it not equally foolish to encourage students to build their life on Christ, the firm foundation, while at the same time teaching them to build their view of math on the sands of worldly thinking? How much better to encourage them to build every part of their life, including their view of math, on Christ!

Return to our sample lesson on division.

Searching for a missing factor is called **division**. A division problem is like a miniature multiplication table. The product is inside the box. The two factors are outside the box. One factor is in front and the other is on top.[26]

This time, instead of adding Bible verses to the lesson, we will try changing the heart in the presentation. Rather than introducing division as an independent fact, we will present division as a practical way of recording a principle God created and sustains. Our division lesson might resemble the following:

Today we are going to learn about division. Suppose you have twelve cookies you want to share evenly amongst three friends. How many cookies should you give each friend? Try it! Take a piece of paper and cut it into twelve circles to represent the cookies. Now split the cookies (circles) into three equal piles. How many cookies are in each pile? Four. You just did division! You split, or divided, a big number (twelve cookies) into a certain number of piles (three) and found out how many went into each pile (four). Guess what? Because God holds all of creation together so consistently, if you take twelve objects and put them into three piles, you will always end up with four objects in each pile! God's consistency makes it possible for us to do division without actually touching the objects we are dividing. Just as we did with multiplication, we can memorize our division facts and be confident they will always work because of God's faithfulness.

Look at the piles of cookies we made. Can you think of a way to express the total number of cookies using multiplication? Think about it. We have 3 piles of 4 each, so 3 x 4, which equals 12.

You will notice our textbook refers to division as finding a missing factor. We could look at what we did with the cookies as a multiplication problem with a factor (a fancy name for the numbers being multiplied) missing. We wanted to find the answer to 3 x ___ = 12. You already know that 3 x 4 equals 12—this is a consistency that, due to God's faithfulness, we can rely upon and memorize. It

---

[26] Hake and Saxon, *Math 65*, 74.

would make sense that if 3 piles of 4 equals 12, then the opposite holds true too: 12 divided, or separated, into 3 piles would give us 4 in each pile, which is exactly what we found when we divided our objects! Whether we call it division or finding a missing factor, division is yet another way to look at and record the orderly way God causes objects to interact.

The above lesson presents division as a useful, real-life tool dependent on God's faithfulness. Note: I would not necessarily use missing factors to present division if your textbook does not—I did it here because if the textbook refers to missing factors, the student would need to really understand the term in order to solve problems in his textbook.

Once we had established the basis of division, we could then go on to teach the "box" method of recording division. We could explain that, over the years, men have employed many different methods to help us describe the way God causes objects to divide. Depending on the student's age, we might even show him some different ways people have written division,[27]

DIFFERENT DIVISION SYMBOLS

μορίον	bhâ	12 / 3	)	÷
•/•	⊓	⌐◦	:	∪

as well as some different division methods employed over the years (see

---

[27] Graph of different division symbols taken from Loop, *Revealing Arithmetic*, 56. Based on symbols from Cajori, *History of Mathematical Notations*, 1:268-271.
Top row, from left to right: Diophantus; Bakhshālī arithmetic; common Hindu method (using modern notation); several mathematicians used this symbol; American/English symbol used today (although many mathematicians have actually used it to represent subtraction!).
Second row, from left to right: A variation on our current symbol used in a Buenos Aires text; symbol used by Gallimard; symbol used by Da Cunha; symbol used by Leibniz; another symbol used by Leibniz. Note: This list is by no means exhaustive!

drawings).[28] We could then help him understand the benefit to having consistency in the way we represent equations (otherwise no one else will understand what we mean!) and walk him through the typical method of representing division employed today.

Do you see how we've now introduced this method of writing division

DIFFERENT DIVISION METHODS

A. Traditional	B. Liber Abaci	C. Repeated Subtraction			
		95	70	45	20
		5	5	5	5
		90	65	40	15
		5	5	5	5
		85	60	35	10
		5	5	5	5
		80	55	30	5
		5	5	5	5
		75	50	25	
		5	5	5	

A. Traditional

$$
\begin{array}{r}
19 \\
5\,\overline{)\,95} \\
50 \\
45 \\
\overline{0}
\end{array}
$$

B. Liber Abaci

$$
\begin{array}{r}
40 \\
95 \\
5 \\
19
\end{array}
$$

D. Speed Mathematics Simplified

$$
\begin{array}{r}
19 \\
5\,\overline{)\,95} \\
4
\end{array}
$$

as just *one* way mathematicians have agreed to record division? This point is very key, as most textbooks present the mechanics of how to add, subtract, multiply, divide, etc., as if the method being presented was *the* way to perform the operation. This subtly gives the wrong impression, leaving the student thinking of the method as an absolute truth in itself rather than as a standardized way to describe a real-life principle God created.

This principle is especially key to keep in mind when introducing the mechanics of a method—such as how to solve a long division problem. Most students learn the long division method without really understanding why the method leads to the right answer. This robs them of really understanding how the technique describes a principle God created. It also teaches them to memorize by rote rather than to really think mathematically.

---

[28] See Appendix D in Loop, *Revealing Arithmetic* for an explanation of each of these methods, as well as of various addition, subtraction, and multiplication methods.

When teaching long division, we could actually walk the student through the process of dividing multi-digit numbers step-by-step, demonstrating it with objects. We could then show him how the division algorithm helps us find these same answers on paper by keeping track of place value for us (see "Understanding the Long Division Method" on page 36).[29] We have now shown the student that math methods, far from being independent *from* God, are completely dependent *on* Him (they record a principle He sustains). We have also helped him learn to think mathematically rather than merely memorize a method.

The above ideas illustrate how division can be presented biblically. As we will explore in depth later, there are many different ways to teach each math concept biblically. But a biblical math presentation will always introduce the concept as a practical tool dependent on God's continual faithfulness. When we approach and teach each concept in this biblical light, math will stop harming our hearts and start reminding us to trust God.

Isaiah 55:8 reads: "'For my thoughts are not your thoughts, neither are your ways my ways,' declares the LORD." In math, as in every other area of life, we have the privilege of laying aside our old thoughts and receiving God's thoughts. How foolish it would be for us to add the Bible to our old thoughts about math! Rather, may we adopt God's thoughts in our math outlook and presentations.

---

[29] For a more detailed explanation, see Loop, *Revealing Arithmetic*, 73-75.

UNDERSTANDING THE LONG DIVISION METHOD

24 Apples

2 Sets of 10 Apples

4 Apples

Divide into two piles.

1 Set of 10 apples
+ 2 apples = 12

Because of the amazing consistency with which God governs creation, we can use paper methods to help us break large numbers down into smaller problems we have already memorized. To understand how the long division method typically taught works, take a look at a simple problem: 24 divided by 2. Is there a way to break this problem down into a series of simpler problems?

Think about what 24 represents—in our place value system, the "2" in "24" stands for 2 sets of 10. What would we get if we divided $\underline{2}$ by $\underline{2}$? We'd get $\underline{1}$. Likewise, if we divide $\underline{2}$ sets of 10 by $\underline{2}$, we get $\underline{1}$ set of 10.

In the long division method, we use this same basic principle to divide on paper. We first divide the "2" in "24," which stands for 2 sets of 10. 2 sets of 10 divided by 2 gives us 1 set of 10. Notice we put the 1 in the ten's column since it stands for 1 set of 10. Then we divide the 4, writing the answer in the one's column. All the 0s, lines, and subtraction (subtracting 2 from 2 and 4 from 4) are just ways to help us make sure we keep the numbers in their proper place and do not forget to divide any part of our original number.

$$\begin{array}{r} 12 \\ 2\overline{)24} \\ \underline{2} \\ 04 \\ \underline{4} \\ 0 \end{array}$$

The long division method is just one way to accurately represent on paper the principle of division God created and sustains. The figure on the preceding page shows several other ways to divide—the *Liber Abaci* (Leonardo Fibonacci) and *Speed Mathematics Simplified* (Edward Stoddard) methods use different means of keeping track of place value; the repeated subtraction method (employed by the ancient Egyptians) breaks down the problem in an entirely different way. All of them accurately describe the same principle of division.

Math Methods from the Past

As you may have already noted, teaching math biblically involves a practical element. We want students to see math as a method of recording the way objects interact in the physical world so they will realize math's utter dependency on God's sustaining hand. Since most of us are so accustomed to viewing math as a paper exercise, the idea of teaching it practically seems foreign to us. I know I used to wonder how math could possibly be taught without endless drills and memorization.

You may be as surprised as I was to learn that math has not always been taught in the current fashion. Two hundred years ago, most textbooks taught math in an incredibly practical way. Look briefly with me at how math was presented. Seeing these historic math methods shows that math does *not* have to be taught through rote drills. Math *can* be taught as a meaningful tool.

Arithmetic Then...

The *Scholars Arithmetic* was a widely used textbook in the early 1800s.[30] The approach implemented in this book can best be termed practical— practical to a degree I had never imagined possible. Anyone reading this book will feel as though they are reading a here-are-the-tools-you-need and here-is-how-to-use-them handbook instead of a math textbook. As the author says:

The *thing needful*, and that which distinguishes the arithmetician, is to know how to proceed by application of these four rules to the solution of any arithmetical question. To afford the scholar this

---

[30] Educational Research Library (National Institute of Education), *Fifteenth to eighteenth century rare books on education: a catalog of the titles held by the Educational Research Library* (Washington: National Institute of Education: For sale by the Supt. Of Docs., U.S. Govt. Print. Off., 1976), 201.

knowledge is the object of all succeeding rules.[31]

"Any arithmetical question" meant not just problems in a math book, but any *real-life scenario* the scholar might encounter. This book taught math as a tool for real life. Using math practically was presented as the main purpose of learning math, not just as an afterthought.

Imagine learning math, not because you had to, but because each and every concept you learned would help you in your daily life. Imagine learning along with subtraction all of the possible uses of subtraction including currency conversion, bookkeeping, and even basic mechanics. Basic arithmetic two hundred years ago was exciting, practical, and purposeful. As history demonstrates, arithmetic *can* be taught as an exciting, practical tool.[32]

Geometry Then...

Two hundred years ago, geometry was not the pile of proofs it has become today. Geometry in the 1700s remained true to its technical Latin meaning, "earth measure." Geometry served as a tool to help men measure and record the earth.

Goedaesia, the geometry book of the time, was actually a surveying manual. Geometric rules did not just help the student prove an irrelevant fact. They helped him find the height of a tree, the area of a field, or the circumference of a lake. Shapes represented real objects and served a real purpose.

Men in the 1700s used geometry on a regular basis. At the time, few skills were more valuable than surveying. Young men, many much younger than today's typical geometry student, taught themselves geometry so that they could survey property. Knowing the shape and features of the

---

[31] Daniel Adams, *The Scholars Arithmetic; or, Federal Accountant. Containing . . . The Whole in A Form and Method Altogether New, for the Ease of the Master and the Greater Progress of the Scholar*, 5th ed. (Leominster, MA: Adams & Wilder, 1804), 57.

[32] I do not mean to imply that textbooks two hundred years ago were perfect. They were not. They had weaknesses. However, they presented each math rule as an incredibly practical tool the student could apply to his own life. Their practical approach serves as a good example.

land was vital to protecting, expanding, and using property wisely. Since most men designed their own homes and gardens, knowing how to survey was an essential skill. Students learned geometry so they could use it in their own lives.

Math textbooks from the 19th century illustrate that we do not have to teach math in a meaningless fashion. Math has been taught practically in the past and can be taught practically today. It *is* possible to entirely change the way we teach math.

> ## NOTE
> Just presenting math practically does not teach math biblically. What a tragedy it would be to know how to use math practically while remaining blind to the One who makes practical math possible! Yet practical math is the natural outcome of a biblical view of math.

THE JOY OF BIBLICAL MATH

Not only is it possible to change the way we view and teach math, but a biblical approach to math is also very rewarding. Briefly journey further back in history with me and take a look at two very different groups of men: the Classical Greek mathematicians (circa 600-300 BC) and the men of the Scientific Revolution (circa 1500-1600 AD). These two groups approached math in radically different ways. Their comparison demonstrates that our thinking about math directs our actions, and that biblical thinking in math brings rewards and immense joy.

We will start with the Classical Greek mathematicians—men such as Pythagoras, Plato, Aristotle, and Euclid. These mathematicians believed logical, mathematical reasoning could determine anything. They placed their entire faith in human intellect and math. They gave the credit for math's incredible consistency to man and math itself. They viewed math as something totally independent from God.

What were the results of this independent perspective of math? The Greek mathematicians, while they made many contributions to math, were never able to fully understand math and why it worked. Instead of using math to develop practical inventions, they spent most of their time trying to explain how the universe originated and operates using math and human reason alone. Math to them was an infallible reference point, an intellectual exercise, and a testimony to man's genius.

To see the full impact of this Greek philosophy, we need to fast-forward in time to the early Christian Era (300-600 AD). During this time, the church began compromising with Greek teaching. The Greek philosophy and approach to math began to infiltrate European culture, leading to many inaccurate ideas and hampering scientific progress.

For example, Greek teaching led to the misguided belief that the sun circled the earth. The Greeks had used math to "prove" that the sun circled the earth. Due to their faith in math and human intellect, the Greeks did not verify this proof with science. Instead of questioning the Greek model of the universe, Europeans twisted Scripture to support this false view.[33] Why? Because Europeans embraced the Greek faith in math's infallibility and man's reasoning. The Greek approach to math and science held Europe in an intellectual and spiritual vise that restricted academic progress and bound society to false ideas.

Moving forward in history, we encounter a radically different group of men: the men of the Scientific Reformation. These men emerged on the scene toward the end of the Middle Ages, a period of time when, despite the corruption of the established church, God's principles began correcting and replacing independent Greek thinking. Most of these men were heavily influenced by the Protestant Reformation, a reformation calling men back to the

---

[33] In an effort to make a geocentric philosophy fit into the Bible, the Medieval church claimed that the sun standing still in Joshua 10:13 meant that the sun circled the earth. However, at the time Scripture was written it was commonly known that the earth circled the sun. Scripture merely uses the earth as a point of reference, which is actually the most scientifically correct method of referring to heavenly movements. See Henry Morris with Henry Morris III, *Many Infallible Proofs: Practical and Useful Evidences for the Christian Faith* (AZ: Master Books, 1996), 253. Quoted in Russell Grigg, "Joshua's Long Day," Creation, June 1997, http://www.answersingenesis.org/creation/v19/i3/longday.asp.

Bible as the basis for every area of life.

Instead of seeing math as an intellectual pursuit of truth, the men of the Scientific Reformation viewed math as a way of recording the truth God had already created and sustains. They embraced math as a vital part of science. Although their theologies varied and were by no means perfect, they generally acknowledged their utter dependence on God and recognized Him, not math, as their ultimate authority.

No longer blindly trusting math and man, men such as Galileo Galilei and Johannes Kepler did not just accept the "proven" geocentric view of the universe. Rather, they questioned the Greek proofs, knowing even the best-laid mathematical proofs could have flaws. Instead of placing their faith in mathematical reasoning, they placed their faith in the Bible. Learning from the sacred pages where the universe originated, these men wasted no time in idle speculation. Instead, they looked in awe at the heavens God stretched out above them and used math to record what they saw. By so doing, they discovered numerous laws (such as the laws of planetary motion) and realized that, contrary to the Greek proofs, the earth circles the sun.

——— NOTE ———

Seeing how a biblical view of math helped bring about the incredible scientific discoveries of the Scientific Reformation should encourage us that the Bible's principles work. We can safely build our understanding of math on God's Word.

Besides just being able to discover previously unknown facts about the universe, these mathematicians/scientists experienced the joy of seeing God's hand as they used math. They knew beyond a shadow of a doubt that math and creation were the Creator's masterpieces and mere reflections of His power. They knew math worked because God faithfully sustains the universe.

For example, Johannes Kepler, the discoverer of the famous laws of planetary motion, clearly acknowledged that the intricate workings of the heavens testified to the Creator's hand. In fact, he realized God had actually hung the planets in such a way that they produced a sound. Kepler saw math as merely showing us design and sounds otherwise hidden from our sight.

If you go back and read his mathematical works, you will find exclamations of praise interjected between explanations of his discoveries. Kepler concludes his famous *Harmonies of the World*, in which, among other things, he shares his third law of planetary motion, with: "Crying out with the royal Psalmist: Great is our Lord and great His virtue and of His wisdom there is no number…To Him be praise, honor, and glory, world without end. Amen."[34]

His biographer states that to Kepler, "God is the beginning and end of scientific research and striving."[35] Regarding the laws of the planets, the

---

[34] Johannes Kepler, *Harmonies of the World*, Book Five. Found in Stephen Hawking, ed., *On the Shoulders of Giants: The Great Works of Physics and Astronomy* (Philadelphia: Running Press Book Publishers, 2002), 1157. Note: Spelling in quote has been updated for clarity; the original read "honour" and "wourld."

[35] Max Caspar, *Kepler*, trans. C. Doris Hellman (New York: Dover Publications, [1959] 1993), 374,

same biographer comments that Kepler believed "the eccentricities of these ellipses are no more arbitrary and without rule than any other measures. No, in this fine construction the highly artistic formative hand of the Creator is shown in a very special way."[36]

Kepler knew why the heavens were orderly. He understood why math worked. The Bible gave him the answers. While his theology and outlook on the universe were by no means perfect, he viewed each scientific and mathematical endeavor as a chance to see more of God's hidden design. He approached math with a heart to see God, and as a result, he was able to make startling discoveries about the universe.

Like Kepler, other men of the Scientific Revolution saw God in math because they approached math biblically. A total heart change enabled them to view and use math appropriately and effectively.[37]

If you want your children to see and use math biblically, I would encourage you to do more than just add Bible verses to your curriculum. Let God change your heart toward math. Like Kepler and the other 16th and 17th century mathematicians who rejected the independent Greek thinking, reject the current independent thinking in today's "neutral" math books. Adopt biblical thinking.

Embracing a biblical philosophy toward math will enable you to use math practically in such a way that math will be a monument in your hearts to God's greatness. As you begin to see and use math biblically yourself, you will be able to teach math biblically to your children so they, too, can behold God in math.

---

quoted in Nickel, 117.

[36] Ibid., 386.

[37] For more information about mathematicians throughout history and their belief systems, see Nickel, *Mathematics: Is God Silent?*

CHAPTER 7
Preparing to Teach Math Biblically

Like anything else in our Christian walk, teaching math biblically will take strength and endurance. It is unnatural for us. We are used to viewing math independently. For that matter, we are used to living independently from God. Drawing our strength from Him and looking at things His way runs contrary to our sinful nature. The Bible says we have to train ourselves for godliness (1 Timothy 4:7). Training implies effort. The rest of this chapter will provide a few training tips.

REMIND YOURSELF OF GOD'S TRUTHS

Continually reminding yourself of God's truths will help you keep on course. "Stand firm then, with the belt of truth buckled around your waist" (Ephesians 6:14a).

Below are a few key truths to keep in mind as you teach.

❖ God created math.
❖ Math works because of God's faithfulness in sustaining all things.
❖ Math helps us explore and utilize creation. As we use math, it should also make us praise the great Creator.
❖ Math is not a neutral subject.
❖ Math was designed to bring glory to God.

If you want to be prepared to teach math biblically, continually remind yourself of God's truths. Surround yourself with as many biblical resources as possible. By whatever means you can, keep God's truths continually before you.

Examine Your Goals

The Bible tells us David's heart was, "fully devoted to the LORD his God" (1 Kings 11:4). We constantly need to ask ourselves if our hearts are fully devoted to God in math.

Many times we devote ourselves to our own goals and desires when we teach. We want to be good teachers. We want our students to succeed academically. We strive for these things and devote our time and energy in these directions. We worry unduly over our students' academic progress.

But stop and think for a moment. What is God's goal for your child?

From the Bible, we learn that God's goal for all His children is for them to walk and live in His Spirit, leaning on Him. The *Shorter Catechism* words it this way: "Man's chief end is to glorify God and enjoy him forever."[38]

The world defines success as good grades and intellectual knowledge, but God does not. God's primary goal in math, as in everything else, is for us to worship and learn from Him. Success in His eyes is seeing and trusting Him. You can teach your children to be proficient in math, but if they have not learned to glorify God in the process, they have fallen short of His goal.

I would like to encourage you to make God's goal your goal in math. Strive to teach your children each math concept biblically. Whether they can manipulate numbers and quote rules should be of secondary importance. More importantly, do they realize how each math concept points to God? Do they know how to use it for His glory? Do they have a firm foundation and perspective in math? Are they applying that perspective to each new concept?

A thousand other goals will continually try to make their way into your heart. When my mom first started homeschooling me, her goal was for me to love God and read His Word. Over the years, however, God began

---

[38] *The Shorter Catechism*, agreed upon by the Reverend Assembly of Divines at Westminster, quoted in *The New England Primer, Improved....* (Boston: printed by Edward Draper, Newbury Street and sold by John Boyle, Marlborough Street, 1777; reprint ed., Aledo, TX: WallBuilder Press, 2003), Question 1.

to show her that she had many other hidden goals for me. She wanted me to be everything she had longed to be herself. She wanted me to succeed and prosper. These other goals and desires often hampered her teaching and kept her from focusing on God's goal. She found she needed to repeatedly surrender and ask God to truly make Him her only goal.

As the Lord shows you other goals you have for your children in math, surrender those goals to Him. Continually ask God to make Him your only goal.

The first step to teaching math biblically, and one to which we need to constantly return, is to make sure our own hearts are truly prepared to worship God in math. We need to constantly keep God's truths and goals in the forefront of our hearts and minds.

CHAPTER 8
Teaching Math Biblically

Every person is so unique and different that no two parents could possibly teach math the exact same way. The way you teach math will be uniquely yours. You may or may not be naturally gifted in math. You may or may not have the same ideas and teaching style I do. You may or may not have time to prepare. Your time with your child in math may be short or long. You may or may not use a textbook. And that is okay! God knows your makeup and limitations. He custom-designed you for your child. So please do not take the ideas I share here as rigid rules to follow. View them instead as guidelines and principles. Let them be a starting point for you, not a formula to follow precisely. Allow God to teach you and give you other ideas.

General Guidelines

While there is no one formula for teaching math biblically, some universal principles always apply. Biblical teaching will always teach the child to view math biblically and use it practically.

To accomplish this, you will need to do more than simply *say* to your child "God created math" or "Math is practical." You will need to *show* your child how each concept testifies to God and teach him how to use it practically. Suppose you did not know how to drive a car. You might need someone to verbally explain how to drive, but eventually you would need someone to put you behind the wheel and actually teach you how to drive!

Likewise, while you should explain to your child that God is in math and that math is usable, do not stop there. Make sure that you are actually teaching your child to view each concept biblically and use it practically.

Teach your child when and how to use each concept.

Mathematics is like a chest of tools: before studying the tools in detail, a good workman should know the object of each, when it is used, how it is used, what it is used for.[39]

Walter W. Sawyer

Students should know and understand math like a workman understands his tools. They should understand where each tool (math concept) came from and why it works. Then they should learn why and how to use that tool (math concept).

In short, whenever you teach math, strive to

❖ show the student how the concept reveals God's character/design,

---

[39] Walter W. Sawyer, *Mathematician's Delight* (Harmondsworth Middlesex: Penguin, 1943), 10, quoted in Nickel, *Mathematics: Is God Silent?* 290.

❖ teach them to really understand how each rule or technique describes a real-life principle God created and sustains, and

❖ equip them to use the concept practically.

To help convey these ideas, you will frequently want to incorporate

❖ the history of math and

❖ the practicality of math.

You may have observed in previous chapters how math's history and practicality help reinforce a biblical view of math. Seeing the way men throughout history have used math reinforces math's usefulness and connection with reality. Looking at the different approaches men have taken to math emphasizes the importance of our beliefs about math. Viewing historical math methods helps students view our current method as a method rather than as the only way to solve a problem. In these and numerous other ways, understanding math's history and practicality helps us view math correctly as a useful, God-given tool. Therefore, you will want to consistently include these elements in your teaching.

You will also want to work toward teaching each concept with understanding. If children are to learn to use math's various tools, they need to understand them, not just memorize them. They need to really grasp how the rule or technique describes a real-life principle God created and sustains so they learn math, not as a set of independent facts, but as a useful tool resting upon God's faithfulness.

One of the biggest challenges in teaching math is to not lose sight of the *principle* in all the terms, rules, and methods. Your goal as a teacher is to show your student terms, rules, and methods in such a way that he understands how the method or rule works and how other methods or rules could also work. The methods vary; the principles do not.

With this general overview in the back of your mind, let us look in more depth at specific ways you can teach your child to view math biblically and use math practically.

> ------ NOTE ------
> While some of the examples below might not be age-appropriate for your child, you can modify them to fit your child's math level. It is never too early or too late to begin teaching math biblically!

View Math Biblically

How can you teach your child to view math biblically? As mentioned in previous chapters, start by approaching math with a heart dependent on the Lord and His Word. Then, look at every concept you teach and ask God to show you how He views it. Much as we did with the equation "1 + 1 = 2" in the first chapter, strive to present the concept in a biblical way. To illustrate, I have listed a few general math concepts below and summarized how each concept can be presented biblically.

Counting—Counting should be introduced as an extremely useful tool that works because of God's faithfulness in holding all things together. In a very basic way, counting helps us quantify the order God has placed around us.

For example, count your fingers. Now count someone else's fingers. Of course, you both have five fingers on each hand. Although we do not often think about it in this light, everyone has five fingers because the same God created us all, and He decided that hands would have five fingers.[40] Counting should make us praise our Creator.

As you teach counting, explain to your child that the Bible also tells us why we can count, or for that matter, use math at all. We, unlike the animals, can count because God created us in His image, capable of seeing and recording the order around us.[41]

When it comes time to teach written numbers and place value, help your child view written numbers as just one way to "name" the quantities

---

[40] Exceptions to this exist because of the fall of man. The world is no longer perfect like God created it, although we can still see remnants of its original order.

[41] Animals were not created in God's image. Thus it is impossible for them to reason in the same way man can. For more information on this or other issues dealing with creation, visit www.answersin-genesis.org.

around us.

Division—When you teach division (or other basic operations), remember why division works (because God holds all things together). Present the many different ways of writing division problems ($y \overline{\smash{\big)} \frac{x}{z}}$, $\div$, $\frac{x}{y}$) as different methods men have agreed to record the fixed law of division God has established and sustains. Help your child understand how each of the different methods he learns actually does describe a real-life principle, so he will learn with understanding and view the method as a *tool* rather than as a *truth* in itself.

Algebra—Since algebra integrates numbers and letters, stress that those letters can represent real-life objects or ideas. For example, the letters x, y, z, etc., could represent distance, apples, the number of items needed, or any other unknown value. By using letters instead of numbers, we can write general relationships (such as the relationship between the area and sides of a rectangle, $a = l \times w$—area equals length times width) that hold true no matter what numbers we insert into the equation.

Algebra is a helpful tool because it enables us to describe, not just one particular situation, but the way things will interact in *any* situation. Why do things interact so consistently? Because God faithfully holds creation together in a consistent manner. Algebra is a beautiful testimony to God's faithfulness and power!

Much of algebra deals with conventions—agreed upon ways of writing operations and equations. Men have not always used these conventions—for years, men wrote math problems out in words. Later, each math book used a different set of symbols! Writing things the same way saves us a lot of time and energy.

Geometry—Geometry means "earth measure." It's the branch of math that helps us measure the earth—from distances in outer space to the microscopic. Geometry helps us lay out gardens, create maps, build houses and machines, design ships, survey, find the height of a mountain, understand the motion of heavenly bodies, explore sounds, develop 3-D simulations and graphics, graph complex statistical data—the list could go on and on. Make sure you present geometry as a useful tool.

When teaching proofs (a key part of geometry), show your child their purpose. While we can observe that the angles in various triangles we might draw all add up to 180 degrees, a proof lets us show that that's always the case when we start with the assumptions in Euclidean geometry (the geometry typically taught in high school). That way, we can use that knowledge to help us measure angles in triangles we can't physically measure—such as a triangle formed by connecting the location of a planet with two other locations. Proofs also help us build critical thinking skills, which can help us discern through arguments in other areas of life, as well as think through problems. Show your child, though, that our very ability to think critically is dependent on God having made us in His image and given us that ability—and what we conclude in a proof is only as valid as the assumptions we make. We can prove something with a valid proof that is not true in the end due to faulty assumptions that were made. A geometry course provides an excellent opportunity to really teach logical thinking and how to discern through arguments.

Use Math Practically

Telling a child the biblical view of a concept is one thing; showing him how to use the concept practically so he will actually understand how it describes God's creation and be able to use it to serve Him is quite another. Here are some ideas to get you started.

Lay aside the textbook as often as possible and have your child apply math to real-life situations. Beyond just helping him learn math practically, using math in his own life will reinforce for your child that math, far from being a man-made system, is a useful way of recording and utilizing principles God created and sustains. Real-life situations help emphasize the fact that math is more than a mere intellectual exercise and equip the child to use man math to serve the Lord.

There are many ways that you can incorporate real-life situations into your math teaching. Below are a few examples. While the examples are geared toward the younger grade levels, the same principles apply in the older grades.

Sorting wash or setting the table can demonstrate the function of sets. You can have your child learn and apply basic addition, subtraction, multiplication, and division with real objects, such as pencils, toy cars, dry elbow noodles, beans, candy, or anything else you have on hand. While teaching graphs or measurements, you can have your child measure his different body parts (such as his arms, wrists, and legs) and cut a piece of ribbon the size of each part. Then have him use the ribbon to graph and compare the sizes of his different body parts. When presenting decimals, you can open a play store and let your child operate the checkouts (he will not even guess that he is learning math). You can incorporate real life into fractions by having your child double or half a recipe with you, or by going on a made-up treasure hunt around the house and dividing the spoils equally.

These examples are just a few of the thousands of simple and fun ways you can teach math without a textbook. Remember, bringing math into real life can help you show your child how math, far from being an isolated, intellectual subject, is a useful, God-given tool.

Sometimes reinforcing a concept with a real-life example is not enough. It is often better to demonstrate the concept in the physical world *before* showing the child how to write and work with it in the textbook. This teaches him to think mathematically and gives him a proper perspective on the concept when he later learns it in his textbook. Above all, it reinforces that math is a way of describing the physical world.

Instead of always presenting concepts to your child, try to occasionally let him discover a concept on his own (with your guidance, of course). Let your child investigate real-life problems and learn to use math to solve those problems first, and then teach him the textbook mechanics of the concept. Your child may even think of his own way to represent what he's discovered—and that's a great thing!

For example, instead of reading to your child about sequences, look at several plants that contain numbers in the Fibonacci sequences (the leaf arrangement in many plants follow this sequence—you can consult a reference book or the Internet to find out what materials to have on hand). Then guide your child into seeing for himself the unique sequences God used in creation. You have now taught him how to explore God's world with math.

Suppose you are teaching rounding. Instead of reading a textbook, take your child to the grocery store and tell him to keep track of all the items you put in the cart. Your child will quickly realize that there is no way he can ever keep track of $1.23 + $2.99 + $2.50, etc. Gently guide him into discovering rounding for himself as you shop. Not only have you taught him rounding, but you have reinforced in his mind math's usefulness. He has learned how to round numbers in a real situation *before* he ever learned rounding in a textbook.

One mom shared with me how she adapted this idea to fit her family. She told her oldest daughter she would buy her and all her younger siblings a piece of candy if she could estimate their groceries for the week within two dollars. With the prospect of candy as an incentive, the daughter paid careful attention the entire time they shopped and actually estimated the total within a few cents. Amazed, the daughter later remarked, "Wow, Mom, math really *does* work!" The younger children, who had watched the whole proceeding with a vested interest, also learned that math could be fun and useful.

Teaching math from real-life situations works. It helps the child view math correctly as a God-given, practical tool. While I do not mean to imply that you will *always* teach without a textbook, I would like to encourage you to break from the idea that you must *always* teach from a textbook. Since we want to see math as a real-life tool, we need to remember to use and teach math from real-life situations.

PUTTING IT ALL TOGETHER

To give you a better idea how the above guidelines impact the way you teach, I would like to walk through teaching two concepts—lines and fractions. First assume that your textbook is presenting the definition of a line. It will probably say something similar to, "a line is a never-ending series of dots," and then move on to defining line segments or number lines. How can you apply the principles mentioned above to this lesson so that your child will view lines biblically and use them practically?

You could rework a lesson on lines in many different ways. How you choose to present lines will depend upon the ideas the Lord has given you at the time. Some ways might require research on your part, but do not panic: the next chapter will explain how to easily find the information and ideas you need.

One way to teach lines would be to revise the textbook's independent presentation so that it presents lines biblically. For example, instead of saying, "a line is a never-ending series of dots," you could explain that a man named Euclid decided to name a never-ending series of dots a line. This little change helps present a line as a *name* chosen by a mathematician to express a real-life concept, whereas the original wording gives the impression that a line is some sort of *self-existent fact*.

Or you could use lines to discuss the vastness of our universe with your child. To us, lines are a never-ending series of dots. We could not even begin to draw a line that would reach the end of our universe. Yet God holds the whole universe in His hands. He knows its end and beginning. Lines are a beautiful testimony to God's mightiness! Since a ruler is really a portion of a number line (or a line segment), you could incorporate measuring into the lesson. Select a project involving measuring so your child can use number lines himself.

Lines also provide a great opportunity for you to discuss the Greek mathematicians. You could share how the Greeks failed to see math biblically and the harm this caused.

If you have time to do a little more research, you could look up lines in a reference book. There you will discover many other definitions for a line. Most textbooks define a line as a never-ending series of dots. This is the definition of a line in Euclidian geometry. Riemannian geometry, on the other hand, gives a line an entirely different and contradictory definition.[42] Both definitions are useful in different situations.

---

[42] Riemannian geometry, in simple terms, defines a line as a "great circle." This would mean a line has a beginning and end and that the equator is really a line, since it has a beginning and an end. See Nickel, *Mathematics: Is God Silent?* 178-179.

Introducing these contradictory definitions to your child is a wonderful way to teach lines. You could explain that, just as forks and spoons help us eat different foods, different geometries help us in different situations. A fork has its limits on what it can help us eat (it does not do well with soup), and geometry definitions have limits on what they can help us record. God has created a complex universe, and each of the geometries helps us explore and record a portion of it. None of the geometries contain perfect definitions—they are merely tools that must be used appropriately to be accurate.

Another way you could teach lines, especially to older children, involves setting aside the textbook entirely. Instead of reading the lesson, you could have your child research lines in other math books and report to you on their meaning, history, and uses. This way, your child will remember the concept *and* learn to view and use lines appropriately.

As you can see, lines can be taught in many different ways. Notice how each way reinforces a biblical understanding of math. Much as there are many different ways to explain or illustrate the gospel message without changing the message, there are many different ways to teach math concepts. Yet each way should reinforce a biblical understanding of math.

Now suppose your textbook is teaching fractions. It will probably go into depth about numerators and denominators, what they mean, and the proper way to write them. After briefly introducing the mechanics of fractions, most curriculums make students solve numerous problems. Instead of merely making your child solve a host of rote problems, how can you show your child a biblical view of fractions and teach them to use fractions practically?

To help your child develop the correct view of fractions and the way to write them, you could explore with him how different cultures have written fractions (for instance, the Egyptians wrote ½ as ⌒).[43] Seeing other ways of writing fractions will help your child view our current way of writing fractions, much like our current language system, as *one* way of expressing a real-life concept.

---

[43] Adler, *Giant Golden Book*, 43.

A helpful point to share with your child would be to explain how fractions represent both partial quantities *and* division. The word fraction actually has its root in the Latin word *frangere*, meaning "to break."[44] You could pull out an orange and ask your child to suppose you wanted to break, or divide, it among four people. Each person would end up with ¼ of the orange. We could look at the ¼ here two different ways: as 1 part out of 4 or as 1 orange divided by 4 people. Understanding fractions as division helps make sense out of the rules to add, subtract, multiply, and divide fractions.[45]

You could also teach fractions by incorporating some science into your math lesson. Lots of simple science experiments involve measuring or weighing fractional amounts. Performing a science experiment can relieve the boredom of studying fractions and demonstrate to your child many practical uses.

Or say you love music. You could draw some music notes and explain how we represent the lengths of time to hold a note using whole notes, half notes, quarter notes, eight notes, etc. As you might suspect, the half note is worth ½ of a whole note. Musical notes are a simple example of fractions in action.

On other days, you could use the textbook a little more. Assuming your child understands fractions biblically, you might decide to let him read and work problems in his textbook for math class. Then, to reinforce that numerators and denominators represent real things, you might have him cut the pie for dinner that night into eighths, or divide objects around the house into fractions with you.

Do you see how the above ways of teaching both lines and fractions followed the guidelines we discussed earlier? They presented lines and fractions from a biblical view. Many of them provided opportunities for

---

[44] *The American Heritage Dictionary of the English Language*, 1980 New College Edition, s.v. "fraction."

[45] See Loop, *Revealing Arithmetic*, 79, for a further explanation.

the child to understand and use the concept away from the textbook. Others brought history and science into the lesson. All of them left the child viewing the concept as *one* tool we use to describe God's creation.

As you teach math, strive to use whatever concept you are teaching as an opportunity to show your child how math is dependent on God and how to use math appropriately. This, in summary, is what teaching math biblically is all about.

CHAPTER 9
Ready, Set, Now What Do I Do?

You sit down to plan your child's math lesson. You open the book and stare blankly at the page. You have no idea how this concept should be taught biblically. What do you do? How do you find ideas?

Below are a few suggestions to help you when you feel at a loss. First, write down what concept(s) you will be teaching to your child (sometimes this simply involves copying the Table of Contents). Then ask yourself the following questions:

❖ How does this concept testify to God's faithfulness or power?
❖ How does this method or rule describe a real-life principle God created and sustains? Is this method or rule a way of keeping track of place value, a standardized way of writing a quantity, etc.? What real-life principle does it describe?
❖ Why is it important to learn this concept? What can be done with the information learned? How is this concept used?
❖ How has this concept been used in the past? What is its history? Did the men who discovered this concept use math correctly? Why or why not?

Asking these questions can prove helpful in seeing how the concept points to God and serves as a practical tool.[46] If you do not know the an-

---

[46] Notice how these questions directly relate to the general guidelines for teaching math on page 49-51. These questions help us think of how the concept reveals God's character/design, describes a real-life principle, applies practically, and relates to science and history.

swers, do not worry. Sometimes God's hand in a concept is not obvious. Sometimes it is not clear how a rule or method works. Oftentimes you may not know why the concept is important to learn or how it should be used. You frequently may not know a thing about the concept's history. But there is no need to panic; as you dig a little deeper, the answers to many of the questions will emerge.

View math as a treasure hunt. Sometimes a treasure is easy to find; other times it takes a little more effort. Likewise, ways to teach a concept biblically will sometimes be obvious to you, while at other times you will have to hunt for the answers you need. Much as a miner who spends his whole life digging for gold, embark on a life-long discovery trip. If you are constantly on the lookout for ways math points to God and is useful, you will find lots of ideas for teaching.

You may find it helpful to create an idea notebook to aid you while you teach. Use this notebook to jot down all the practical uses for math you discover. If you find yourself using math to measure fabric for a dress, write that down in your notebook. If you read about a new scientific discovery in the newspaper that used math, make a note of it. You can even bring your children in on the hunt, having them write down any uses they notice. Then on those days when you need ideas, you can use your notebook for inspiration.

In appendix A, you will find a sample notebook. The notebook lists a few common ways almost all of us use math, as well as some basic ways you will likely encounter math in science. The examples listed there are just to get you started. The rest of this chapter will show you how to begin to find your own ideas for math in everything you do.

In Your Curriculum

Start hunting for ideas in the math textbook you are using. Check to see if it gives a clue on how the concept was discovered or used. If it does, take the examples it offers and explore them in more depth with your child.

In Resources

It is often helpful to consult other resources besides your math textbook. Several good resources are listed on pages 78-83. You may want to have a few of these resources on hand.

Also, take advantage of the resources at your local library. For instance, if you need graphing ideas, you should easily be able to locate a helpful book on graphing in your library's junior section. Look for books with key words such as "practical" or "real life" in their title or description. I have found several useful books by just scanning my local library's bookshelves.

The Internet is also a great information source. Frequently, you can learn more about a math concept simply by typing its name into a search engine. For example, I found details about the mathematical way seeds are arranged in sunflowers by searching for "Fibonacci sequence" online (I learned of the existence of this sequence initially by looking up "sequences" in the index of *Mathematics: Is God Silent?*, one of the reference books listed in the next chapter).

I would also invite you to explore the links and resources I have posted on my math blog (www.christianperspective.net/blog) and to subscribe there for updates of new posts. You can also contact me with questions and ideas there (I welcome your correspondence).

In Your Own Life

For other ideas, think about how you have used the concept in the past. We all use math a lot more than we realize. Many times, the ways you have used a concept will give you ideas on how to teach it as a practical tool.

Keep your eyes open for math in other subjects. Since science is particularly closely connected with math, you can often find ideas by seeing if math was used to help discover or is used to apply the science concept you are studying. For example, while studying earthquakes, research how math is used to record seismic activity. You will generally discover practical uses for several math concepts in each topic you research.

History provides a wonderful opportunity for you to explore math's past uses. For example, math enabled the Egyptians to build gigantic pyramids. So when you study Egypt in history, why not have your child build a sugar cube pyramid?[47] Use the opportunity to show him the math behind pyramids while reinforcing in his mind math's usefulness and purpose.

When studying Greece, research how the Greek philosophers viewed math inappropriately. As you teach the Protestant Reformation period, explore the lives and math of men such as Johannes Kepler or Sir Isaac Newton. Do not just read about history-making inventions—study the inventions and see how the inventor put math to a practical use. Each and every time period in history can help you teach math.

Music class also provides wonderful lessons on math. Good music follows an orderly and mathematical structure. Whenever you listen to music, pay attention to the song's order or disorder. In the eighteenth and nineteenth centuries, composers built their songs around tones that were mathematically similar. This logical selection of notes created pieces that sounded beautiful and soothing to the ear. However, many musicians in the early twentieth century stopped using notes that went together mathematically. As a result, their music sounds strangely unconnected.[48]

Even art class can turn into a math lesson. After all, good drawings contain proper proportions and shapes. Paints and other mediums must

---

[47] Instructions on building sugar cube pyramids can be found in Cindy A. Littlefield, *Real-World Math for Hands-On Fun!* (Charlotte, VT: Williamson Publishing, 2001).

[48] Any guide to classical music should explain the general regularity, or lack thereof, in a piece. Pay close attention to the word, "dissonance," a fancy word for notes not mathematically related to each other.

also be mixed in precise mathematical proportions to produce the desired effect.

Wherever you are and whatever you are doing, have a learning attitude. Be ready for God to show you something. Continually hunt for God's ideas in everything you do. Make math a daily treasure hunt. Then you will be ready with plenty of ideas to incorporate into your math lessons.

You do not necessarily need to have each concept researched and understood before you start teaching. You can let your child help you through the process. Math can be a fun treasure hunt for both you and your child.

CHAPTER 10

Curriculums and Resources

When I wrote the first edition of this book, there was no curriculum out there that really taught math in the way I've outlined. Since then, God ended up leading me to develop curriculum and further resources in order to make it easy for parents and teachers to teach math from a biblical worldview. So if you're looking for ready-to-go material you can use that implements the principles presented in this book, please check out what's available at ChristianPerspective.net.

But for the grades I have not written yet or those wanting to use a different program, I want to take this chapter to explore both what to look for in a curriculum as well as how to go about modifying a curriculum, especially one that does not necessarily present math from a biblical worldview. I will also touch on a few other curriculums/resources currently available. Please note that the curriculum list is by no means comprehensive; it consists of ones I've had an opportunity to personally look at. Please see my website for further updates.

What to Look For in a Curriculum/ How to Modify Your Curriculum

When choosing a curriculum, keep your goal in mind. Above all, you want your child see math as God sees it. You want him to understand how math symbols and methods represent principles God created and sustains. You want to demonstrate this through everyday, historical, and scientific examples and exposure.

Ideally, you want to find a curriculum that presents math biblically, but remember, the word "Christian" on a math book does not mean it presents math biblically. There are many "Christian" textbooks that look identical to a secular textbook except for a Scripture verse or saying in the corner. A truly Christian textbook is one that recognizes God's hand in math and presents math in a biblical and practical way.

Since many Christians believe math is spiritually neutral, there are few biblical math curriculums available. The good news, however, is that there are some, and that there has been a resurgence of practical math curriculums in recent years, as well as curriculums designed to teach with understanding. The more practical, scientific, and historical information your curriculum contains, the better. Your curriculum should present math as a tool rather than a set of rules to memorize. Look for a curriculum that

- ❖ teaches concepts with understanding, explaining how the different rules work rather than teaching them by rote (thereby helping the student see how concepts describe a real-life consistency);
- ❖ makes the student use math practically (preferably not with meaningless word problems but in conjunction with science, accounting, mechanics, or other disciplines);
- ❖ illustrates how concepts are used in science and other disciplines; and
- ❖ gives historical information on concepts.

Each curriculum has strengths and weaknesses; whichever of the elements above your curriculum is weak in will be the element you will want to add more of to the program.

Factors such as cost, usability, and learning styles may also impact your curriculum choice. Some curriculums, while they might work great for one student, would frustrate another one. Some curriculums proceed at a faster pace than others—a faster-moving program might not work well for a student who does not pick up concepts quickly. Some courses present the big picture of concepts, then delve into the details, while others just start with the details—a detail-oriented course might lose the "big picture" type child.

Likewise, some curriculums demand more teacher involvement than others. These higher-involvement curriculums, while excellent for some families, might not be the wisest choice for a parent already pressed for time. Some curriculums are also easier to use in a classroom than others. The curriculum you select will depend in a large measure on your specific needs.

In fact, you may find it easier to not use an official curriculum at all. Instead, you may choose to assemble your own curriculum from various resources. There are numerous books on math that, while not actually curriculums, contain a lot more practical and historical information than most curriculums. Do not be afraid to supplement or even replace official curriculums with these books.

If you do opt to use an official curriculum, do not feel obligated to follow the curriculum exactly. View it as a guide to help you, but do not let it stifle you. Remember, curriculums are written for a variety of students, not your particular child. Curriculums have to include a lot of problems because some students may need to work a lot of problems in order to master the material. This does not mean you necessarily need to make *your* child work every problem in the book; he may only need to work a few if he really understands the concept. As you supplement the curriculum with real-life applications, you may find you do not need to assign as much of the textbook work.

If you find your child is struggling in math, ask God for wisdom as to why and how you can go about changing things to help. Does he understand the concepts? Is there a concept that the concept your looking at builds on that he didn't understand or forgot? For example, a student struggling with fractions in algebra may not have completely grasped fractions back in elementary/junior high and may simply need to review the concept with known values before attempting it with unknowns. Did the child just memorize a formula without really understanding the principle? While memorizing step-by-step processes has its place, it's important that students understand how those processes are really keeping track of real-life principles God created and sustains. A big part of math is learning to think through how to actually explore God's creation and complete real-life tasks. A student who hasn't been taught math this way, though, may struggle when put in a curriculum that has a lot of real-life application. They'll simply need some extra guidance at first in understanding how to really use math as a tool. Is the child just being careless? Consider insentivizing them to take care by limiting the number of problems they have to solve, adding the remainder only if they get ones wrong. Again, remember that a curriculum is a guide, not something set in stone. Feel free to adapt as needed for your child.

------ NOTE ------

Just because a resource is listed in the following sections does not mean I recommend or endorse it. I have listed a few of the different curriculums currently available, and a handful of those reference books I personally found helpful or thought would be helpful to you due to the mathematical information they contain. I do not agree with all of the spiritual content in all of the listed sources, so please be careful to hold everything up to scriptural truths. I have based the reviews off looking at a portion of the curriculum or resource; I have not examined every grade level of every curriculum or every section of each resource.

CURRICULUMS

The curriculums are listed in alphabetical order. The length of the description does not in any way reflect a preference of one curriculum over another. It is my intent to share a few thoughts on a sampling of curriculums so you have an idea of some different approaches available.[49] I have marked curriculum where there are videos available to teach the lessons to the student (marked with a ⬤); note that in some of these programs

---

[49] In order to review some of these curriculum and resources, I may have requested and received a complimentary copy from the publisher; however, the reviews reflect my own observations and thoughts.

the video portion is optional, while others are structured around its use.

While almost all programs use manipulatives in the younger grades, a few programs are extremely manipulative based. I've marked these with a ✋. If you use a manipulative-based math curriculum, be sure to also bring in real-life situations. It is important for the student to use math beyond manipulatives. Additionally, remember to add math's scientific uses and historical roles into the curriculum.

Some curriculums contain a lot more written instructions and include detailed lessons for the teacher in a seperate book. I've marked these with a 📖. Homeschoolers using these programs should be aware that the teacher instructions may be more than they need; Christian schools using a program not marked with this should be aware that they may not have the same level of teacher instruction/classroom layout as others.

No matter what curriculum you choose, remember to guard against independent thinking. Reinterpret the math facts through a biblical, dependent philosophy, much like we did with the division lesson on pages 32-34.

Unless otherwise indicated, most of the curriculums do not stress math's history or scientific applications. You'll also notice that, unless where indicated otherwise, many programs don't show math's practicality. Those are all things you'll want to add to the program (or look for one that does incorporate them).

📖 ⚪ *A Beka Arithmetic*—(www.abeka.com) A Beka offers a complete series of full-color textbooks written by Judy Howe. These textbooks predominately follow a drill approach to teaching math. They present math rules and facts, then drill the students on these rules through numerous problems. There are optional videos that show a classroom lesson.

Bob Jones—(www.bjup.com) This textbook series offers explanations of the concepts in the student's textbook, as well as a teacher's edition that gives detailed lesson plans and suggestions (along with manipulative ideas). The authors recognize that math is not neutral and, in many grades, include a profile of math in action at the beginning of each chapter. There is an option to subscribe for an online video course or buy DVDs rather than using the textbooks only.

Horizons Math—(www.aop.com) *Horizons Math* (produced by Alpha Omega Publications) seeks to combine analytical reasoning, hands-on learning, memorization, and drill. On some days, the program incorporates lots of manipulatives into the program, while on other days it focuses on textbook drills and memorization techniques. The teacher presents the lesson, after which the student fills in a worksheet.

Jacob's Math—(www.masterbooks.com) Harold R. Jacobs has written an excellent textbook series for high school students. This series incorporates history and teaches the student to use math in problem-solving situations. This curriculum deals with concepts in a very rigorous way, which may be challenging for some students, especially if they came from a program that did not emphasize problem solving. I have put together an eCourse to go alongside this program that both helps students master the material and adds biblical worldview content.

Life of Fred—(www.stanleyschmidt.com) This series presents math (elementary through calculus, statistics, and linear algebra) as part of an interesting (though purposefully unbelievable) story of a childhood math genius, Fred Gauss. Students read the story, in which Fred invariably uses math, then work a few problems before they are allowed to move on to the next part of the story. The series does a great job connecting math to practical uses and incorporating problems that make students think rather than merely crunch numbers. But don't expect an organized, typical textbook. Note that there are elements in the story line you might wish to discuss with your child.

Key Curriculum Press—Key Curriculum Press offers several different sets of math curriculums that incorporate a great deal of pertinent information. This series does not purport to be Christian; however, they recognize that students need to learn math as something useful and, therefore, include a great deal of historical and practical information.

☙ Key Curriculum's ***Miquon Math*** (www.miquonmath.com) is designed for elementary children. This curriculum includes lots of science and history in a format easy to adapt to a biblical worldview, although it requires a good deal of parental involvement since the teacher's key lacks answers to all but the more challenging problems.

Key Curriculum Press also offers a ***Key to*** series for late elementary and middle school students, which, though laid out in an entirely different format from *Miquon Math,* maintains some of the same practical information. This series introduces everything in the student text and includes a teacher's key containing answers to every problem, keeping parental involvement minimal. Instead of coming in one large textbook, the *Key to* series comes in sets of smaller workbooks. Each set of workbooks covers a specific concept, such as fractions or measurements. This series was designed to supplement other curriculums, although some may find it also works well by itself.

Math Mammoth—(www.mammothmath.com) Math Mammoth offers a series of very flexible textbooks for grades 1-6. The material is available in either a digital format (either downloadable or on a CD) or a printed format. The program includes a lot of worksheets containing drills, pictures, and some word problems, along with simple presentations, game suggestions, and Internet links. It is available both by grade (the Light Blue Series) or by concept (the Blue Series). Free videos are available online. Note that the website does advertise it as meeting and exceeding Common Core.

🖐 ***Making Math Meaningful***—(www.cornerstonecurriculum.com) This K-algebra curriculum does a great job teaching the student how to think mathematically and to understand what is really happening in math. It incorporates a fair amount of manipulative use as well as some textbook work. In the younger grades, the teacher's key tells the parent exactly what to say and do in each lesson. You might find this detailed instruction helpful, but do not let it inhibit you from incorporating your own ideas.

Mastery Publications—(www.masterypublications.com) Mastery Publications offers an affordable series for K through about 6th or 7th grade. Parents are allowed to copy the workbook pages for several children, and all the grades share one parent's manual that offers real-life activity ideas you can use to reinforce concepts. This program primarily focuses, however, on the memorization of math facts and skills, so you will need to be extra careful not to just present math as rote rules and skills.

Math Lessons for a Living Education—
(www.masterbooks.com) This program, which is very simple and intuitive to use for homeschool families, does a good job showing math in action and comes from a young-earth creationist perspective. Students read or listen to a story about two children who discover different math concepts in real life, and then work worksheets related to that story. The stories focus around a Christian family and illustrate life lessons and good character along the way. Students are instructed to make various manipulatives at times. Grades preK-6 are available. Grade 6 was designed to transition into the *Principles* program I wrote for junior high.

Math on the Level—(www.mathonthelevel.com) This truly one-of-a-kind program is based around teaching math as a part of life. Written by a homeschool mom I know, it shows parents how to teach math throughout the day and offers a unique "5-A-Day" system by which parents can track their students' progress and assign daily review problems targeted to their weak areas. They now have an online generator available to easily generate these review problems. The program integrates math with life amazingly well. The flexible structure would make it very easy to adapt. The program is designed for grades K-8 and can be used with multiple children.

Math-U-See—(www.mathusee.com) This curriculum uses base 10 stacking-block manipulatives and pictures to demonstrate the meaning of written numbers and operations. It comes with an onlinepresentation of each lesson. The student builds everything he learns in math with manipulatives. Math-U-See is based on having the student build, write, and verbalize each math concept. It stresses place value and seeks to help the student learn with understanding. Parents will want to add in more real-life applications beyond the stacking block manipulatives.

Mott Media—(www.mottmedia.com) Mott Media has reproduced *Ray's Arithmetic*, a popular math series from the 1800s. The publishers also offer a *Parent-Teacher's Guide* by Ruth Beechick, which shows today's parents/teachers how they can use the series. The guide divides the book into weekly lesson plans, offering ideas on games and ways to teach math outside of the book. The publisher recognizes that math is not neutral. Do not expect, however, to just hand your child the book and tell him to complete a certain number of pages. This program is designed to guide the *parent* in teaching math. The *Ray's Arithmetic* series is also available for free download on Google books (www.google.com/books), although the helpful teacher's guide is not.

Principles of Series—(www.masterbooks.com) This is a junior high and Algebra 2 series I wrote that teaches math concepts from a biblical worldview, incorporating science, history, and real-life examples. It's designed for students to use predominately independently, with the *Student Textbook* teaching the material (or you can get an optional eCourse where students can watch a video presenting the lessons), and then students completing a worksheet. The worksheets include a lot of practical problems and tend to not have as many problems as other programs (but the problems there really stress understanding and application, with the goal of mastering the material without busywork). In Algebra 2, the solutions are in a seperate book that includes detailed notes on how answers were obtained.

Professor B's Power Mathematics—(www.profb.com) The main focus of this curriculum is to help students learn math based on understanding instead of rote memorization. The author recognized that God created us as intelligent beings, and therefore sought to teach math in an intelligent, understandable fashion. The curriculum focuses on arithmetic exclusively—math concepts besides arithmetic (such as geometry) would need taught separately, either at a later date or simultaneously. There is a computer-taught version of the program as well as a textbook version. The company also offers an algebra book. Powerpoint slides are sold seperately.

Purposeful Design Publications—(www.purposefuldesign.com) This K-8 program is produced by a division of the Association of Christian Schools International (ACSI) and features an indepth Teacher Edition that walks teachers through the lesson. The lessons do an excellent job presenting math as a practical tool and bringing in a biblical worldview in various ways. There is a manipulative kit for grades K-6 and some manipulative use; the program also incudes other hands-on activities.

RightStart Mathematics—(www.alabacus.com) This innovative program takes a completely different approach to math. Using an abacus that is color-coded in sets of five, *RightStart* teaches students to view numbers in groups of 5s and 10s as the Asians do, giving students a visual framework for understanding concepts and quickly working with large numbers, and truly enforcing understanding rather than memorization. The last book in the series teaches math through geometric concepts, creatively incorporating science, art, and other math applications. The program is arranged by level rather than grade, but it starts at the beginning and goes through approximately fifth to seventh grade (they recommend moving into *VideoText Algebra* while completing their last book). Transitional books are available to help students adapt into the curriculum at any stage along the way. For those wanting to use them as a supplement to another curriculum, *RightStart* also individually sells their abacus, a book of abacus activities and worksheets, and some very ingenious math games. With the exception of their middle school geometric-based book, the *RightStart* program is teacher or parent-directed, which means teachers or parents should expect to do math every day with the child. Optional video classes are available for some levels.

Saxon Math—(www.saxonpublishers.com) See notes on *Saxon Math Homeschool* below, but note they have a seperate program for classroom settings.

Saxon Math Homeschool—(www.saxonpublishers.com) The Saxon math series utilizes a review system to help students retain math concepts. Because of this review approach, the Saxon books jump around a lot between concepts. This layout, as well as other factors, makes presenting each concept biblically particularly challenging. Saxon never claims to be Christian, so, as you might expect, these textbooks present math as an independent fact.

- *Singapore Math's Primary Mathematics*—(www.singaporemath.com) Singapore Math's *Primary Mathematics* (US Edition) is a fairly popular series of math books developed in Singapore in the 1980s. The actual student workbooks are streamlined, containing mainly graphical illustrations and problems rather than lengthy presentations. The accompanying *Home Instructor's Guide* or *Teacher's Guide* offers the teaching part, sometimes incorporating manipulative use. Note that Singapore Math also has other math series, which I did not review.

- *Teaching Textbooks*—(www.teachingtextbooks.com) This computer-based program currently runs on a subscription basis. Students can watch the presentation (an eBook is also available to read) and complete their problems on the computer for instant automatic grading. Be sure to download a trial to see how it works before using, as it's a very different approach than others.

- *VideoTextInteractive*—(www.videotext.com) These algebra and geometry courses build math step-by-step through videos you can either subscribe to watch online or purchase to play in either a DVD player or computer. The courses clearly present math as a language and make it clear terms and symbols are just *one* set of tools to help us solve a problem and help students understand and think through each new concept. Each day, students (parents are encouraged to join in, especially at first) watch a 5-10 minute lesson (which can take longer, as students are encouraged to pause at each question and think through the answer), then review the provided lesson notes, work several exercises, and check their own work using the solutions manual. Tests are provided every few lessons.

The list above by no means includes every math curriculum available. For additional curriculum thoughts, especially on high school curriculums, see James Nickel's recommendations in *Mathematics: Is God Silent?* Please also see my website, www.christianperspective.net, for any updates there.

Reference Books

Revealing Arithmetic: Math Concepts from a Biblical Worldview by Katherine [Loop] Hannon—(www.christianperspective.net, www.masterbooks.com) I wrote this book specifically in answer to requests from those who had read *Beyond Numbers* and wanted more details they could use in teaching. *Revealing Arithmetic* is a supplemental curriculum designed to help parents teach arithmetic concepts (counting through exponents) to their children (about Pre-K through grade 6, although older students can read sections as a review) in a biblical, exciting way.

Each chapter takes a concept and explains the biblical worldview, offers teaching suggestions, and gives a list of ideas you can choose from as you teach. You will learn the "why" behind the different operations so that you will be able to help your child learn with understanding. You will also find hard-to-locate information on the history of various concepts, ideas on how to transform everyday activities and objects into math lessons, and more—all in a way designed to help you and your child worship the Lord in math. There's also an eCourse available where you can watch me actually walk through the different concepts.

Principles of Mathematics: Biblical Worldview Curriculum by Katherine [Loop] Hannon—(www.christianperspective.net, www.masterbooks.com)

While designed as an actual curriculum (please see its review in the Curriculums section), this program would also serve as an excellent reference tool for anyone teaching math to middle or high school. Both textbooks contain a detailed index, so you can easily look up concepts and read the text presenting them in order to get ideas on how to present those concepts. The Teacher's Guide/Workbook portion can be used to find real-life problems to use with your students.

Mathematics: Is God Silent? by James D. Nickel—
(www.rosshousebooks.com) *Mathematics: Is God Silent?* is a valuable reference book for any math teacher. This extensive book functions like a Christian encyclopedia on math. While you may find this book challenging to read because of its thoroughness and mathematical terminology, it contains a lot of important mathematical information and is well worth the effort. In the first part of the book, you will find concepts and examples intertwined with a complete chronological history of math. The second part of the book explains how numerous math concepts reflect God's character and are used practically. The book also looks at teaching techniques and failures, offers guidelines for teachers, and evaluates a host of resources you can use to teach, especially if you have older children.

When you need more information about a math concept, look up the concept in the index and follow some of the links (try some from the beginning, middle, and end of the book). If the word you are looking for is not listed, try a similar or more general word. For example, when trying to find information about quotients, try looking up division. If you still cannot find what you are looking for, thumb through the book and see if a picture or heading looks related. Familiarizing yourself with this resource will help you find information about a concept quickly and easily.

Truth in the Transcendent by Larry L. Zimmerman—
(https://answersingenesis.org/answers/books/truth-transcendent) This book, currently available to read free online, offers a look at the thought and purpose behind math. It covers mathematical philosophies in technical and historical detail. By the end, the reader has no doubt that the biblical approach to math is the only one that truly makes sense.

Arithmetic for Parents by Ron Aharoni—(www.sumizdat.org) This book shares valuable teaching principles and insights in an understandable way. While the author does not mention the Lord, many of the teaching techniques he has come to are very consistent with a biblical world view (such as explaining why rules work). The book may seem pricey, but I found it well worth the cost.

Maximum Math by Kathryn Stout—(www.designastudy.com/products/mathematics.html) Maximum Math is part of a series designed to help parents and teachers design their own customized studies. It contains helpful listings of sample objectives to use for various grade levels, along with a few suggestions on how to go about teaching those objectives. For example, the kindergarten through grade 2 objectives given include being able to "classify and sort—same kind, color, size, and/or shape."[50] The teaching notes offer general suggestions such as to "match the color of objects or pictures to a color card or place items on a piece of colored construction paper"[51] to help accomplish those objectives. Many of these suggestions are quite clever and provide numerous practical ideas of how to illustrate concepts with real-life objects; the author clearly believes math should go beyond rote memorization. The introductory chapter makes a lot of good points about making children apply and really understand math concepts. Because of its layout and affordable price, this book could easily be used as a reference when looking for further ideas on a specific concept, or as a guide to designing your own study. Note that it is now exclusively in ebook form, although you may be able to find the paperback edition used.

SUPPLEMENTAL BOOKS

There are lots of supplemental math books that, while they do not necessarily claim to be Christian or approach the topic from a biblical viewpoint, contain some ideas you might find useful. Please use your own discernment as to whether a resource is appropriate for your particular family. Your library should also carry a variety of books from which you could glean ideas. Please see www.christianperspective.net/blog for additional resources.

---

[50] Kathryn Stout, *Maximum Math*, rev. ed. (Wilmington, DE: Design-A Study, 1999), 23.
[51] Ibid., 34

Consortium for Mathematics and Its Applications (COMAP) Re-sources—This organization has produced a variety of resources that show math in action. Various ones are often listed on their website for free (www.comap.com/Free). Note that some of their resources may have aspects you will want to discuss with students (in the series *For All Practical Purposes* that I previewed years ago, one or two videos had very brief sections discussing evolution from a non-biblical perspective, but on the whole the videos focused on math's practical uses). Some of their materials can be found used for significantly less. (Note: Be careful when ordering used to make sure you're getting what you mean to, as some materials have related products; you may wish to research the ISBN number listed or confirm with the seller before placing your order.)

Exploring the World of Mathematics **Exploring the World of Mathematics** by John Hudson Tiner does an excellent job sharing some of the history and practicality of math without containing the typical evolutionary bias. Developed in such a way that elementary through high school students can enjoy it, this book works well as a supplement to any math curriculum. Hudson writes in an easy-to-read style (I read it before bed each night), yet at the same time includes a few thought-provoking math problems at the end of each chapter.

Mathematicians Are People, Too: Stories from the Lives of Great Mathematicians by Luetta and Wilbert Reimer—This is an easy-to-read book on the lives of great mathematicians. The authors did an excellent job showing how mathematicians developed many concepts. While the book does not go into depth on the philosophy of the different mathematicians, it touches on it here and there, opening the door for you to explore further and discuss each mathematician's beliefs and practices (and where his or her beliefs led). It is the sort of book I would suggest reading together with your child and discussing as you go.

Math for Every Kid: Easy Activities that Make Learning Fun by Janice Van Cleave —This book combines math and science through a variety of activities.

Math for the Very Young: A Handbook of Activities for Parents and Teachers by Lydia Polonsky & *Games for Math: Playful Ways to Help Your Child* by Peggy Kaye—Like the titles suggest, these books share simple activities to use with your younger math students (preschool to about third grade). Many of these activities are ones you do every day but might not have realized involve math. Some of the activities could also be adapted for use with older children.

How to Lie with Statistics by Darrell Huff—This is by far the best book on how statistics can be twisted I have found. Written in a clear, fun style with plenty of humorous illustrations, you can use the book with upper elementary school on up to adult.

The Only Math Book You'll Ever Need by Stanley Kogelman and Barbara R. Heller—This book explains how to balance checkbooks, calculate interest and loans, determine the yardage for a sewing project, and apply math in other common settings. Although geared for high school to adult level, this book is comparatively easy to read, and many of the concepts could be simplified for elementary or middle school students.

Real-World Math for Hands-On Fun! by Cindy A. Littlefield—This book provides ideas for younger children that integrate math with science, history, and real life.

The Science Book of Numbers by Jack Challoner—This short picture book provides hands-on projects related to math. Instructions are given for making a digital dial, abacus, slide rule, music machine, counting wheel, sundial, and more. These projects would probably work best for younger elementary students with parental supervision.

Science Projects About Math by Robert Gardner—Here is a book that explains the math behind science in a fun and interactive way. Middle school and high school students (and even upper elementary students, with supervision) will enjoy putting math into action with these fun science experiments.

Using Math to... by Gareth Stevens Publishing—*The Using Math to...* series makes math come alive for elementary students by actually allowing the student to use math in different occupations (crime-solving, caring for animals at the zoo, winning a Grand Prix, and others). Students use numerous math concepts in a fun and engaging way.

CHAPTER 11
Overcoming the Difficulties

As with anything worthwhile, teaching math will have its difficult moments. Sometimes you will be faced with personal obstacles or fears. Your child does not understand. You do not think you are good at teaching math. You are confused by the math concept yourself. You have run out of time or ideas. While teaching math may sometimes appear impossible, there are ways to handle the obstacles.

Enormous words and phrases present the most common and largest obstacle to teaching math. You know the words I am talking about. The words you read several times without getting any closer to understanding them. The words that sound so intellectual that you ask yourself what you are doing reading the book in the first place. The words that make you check the title page to make sure you are not reading a Greek or Latin edition.

There have been times when I have read pages in math books that seemed impossible to grasp. I remember struggling for days trying to understand a particular concept in a math book. Once I finally understood it, I could hardly believe I had struggled so hard with such a simple concept.

I like to think of intellectual words as wrapping paper. Wrapping paper often hides useful gifts. Likewise, intellectual words often hide very simple, practical concepts. Remember, every part of math is a practical tool you can use to worship God and explore the earth. Each seemingly complex word you encounter disguises a very simple tool.

Do not be frustrated by initially hard-to-comprehend words. When you encounter a difficult word (or paragraph), stop and ask yourself what tool lies buried in the word. Think back to God's simple truths about math. Think of how they relate to this specific area of math. Remember, the challenging word names some principle God created and sustains, or perhaps names a way mathematicians have agreed to write a problem. Reread the word with God's truths in mind, and you will be amazed at how simple even the most complex word becomes.

As you seek to explore math, you may encounter difficult concepts that, try as you might, neither you nor your child can grasp. Do not panic! When you find yourself lacking knowledge, remember your Heavenly Father is the source of all knowledge. "Wisdom and power are His" (Daniel 2:20). God knows all about math. Let Him teach you and your child.

I remember many times during my childhood when my mother and I could not understand an English concept. Sometimes, we would stare at it without success. Finally, Mom would bow her head and pray, telling God that we did not get it and asking Him to help me understand if I needed to know this concept. Often, after praying and looking at the lesson again, I would understand. Other times it would be made clear to me a few hours, days, or months later. One way or another, God always showed me what I needed to know.

If you do not think you have time to prepare biblical math lessons for your children, consider letting your children prepare the lessons with you. Walk through the errors in the lesson with them and let them help you find more information on how that particular concept should be approached biblically.

I remember my mom and I taking this approach to history lessons one year. No amount of prepared lessons could have taught me so much. Alongside my mom, I learned discernment and research skills. Above all, I watched her pray and ask God's help and strength when we both felt confused. Do not be afraid to learn and grow with your children.

Whether the difficulty you are facing bears the shape of an ominous word, a confusing concept, or a busy schedule, seek God's help and receive from Him the clarity, wisdom, and time you need. Make each new day of teaching exciting by relying on God for ideas and strength.

I still remember the fear I felt on one of my first skiing trips. As I slid off the ski lift and stared down the slope, my whole body started to tremble. I could never make it down this slope! The expert skiers around me called the slope easy, but it looked extremely difficult to me. I pointed my skis slightly downhill. The movement terrified me. I inched forward as slowly as possible. How could people actually enjoy skiing?

As I would later discover myself, people enjoy skiing because they stop focusing on themselves and their fears. They allow their eyes to take in the majestic mountain peaks around them. They breathe in the invigorating mountain air. Instead of letting the enormity of the slope before them overwhelm them, they tackle the mountain little by little. Although they may fall, they have learned to love the snow melting on their face and the challenge of standing up again. They actually relish the moguls that beginning skiers would term impossible obstacles.

Sometimes teaching math may feel as overwhelming as facing a steep slope. Although teaching math at first may seem impossible, do not panic and do not give up when you take a fall. Take your eyes off your fears.

Instead, focus them on the Lord and on the breathtaking view of His character He has displayed in math. Rather than letting the enormity of your task overwhelm you, ask God to help you with the one math concept you need to teach at the moment. Just as I discovered the joys of skiing, you will discover God's joy and blessings in math. The very things about math that once intimidated you will eventually bring you great joy.

Always give yourselves fully to the work of the Lord, because you know that your labor in the Lord is not in vain.

1 Corinthians 15:58

Teaching math biblically is worth all the effort! By showing your child how to approach math biblically, you are showing him how biblical truths affect everything he does. You are allowing him to experience the joy of seeing and using math as a practical tool. You are letting him explore God's creation and magnify his Creator. Above all, you are teaching him the *truth*. Instead of leaving your child with an independent understanding of math, you are letting math become a testimony to God in his heart. Do not be discouraged. Your labor in the Lord is *not* in vain.

PARTING THOUGHT

Exploring math with you has been a blessing. I have found myself challenged and encouraged by God's presence in math—challenged to trust Him more and encouraged that He will indeed be faithful to His Word. Yet I know we have really only begun to see the ways math points us to God. Likewise, we have only touched on the teaching possibilities and opportunities buried within math. My prayer for all of us is that we would be encouraged to further pursue God in math and that in math class, as well as in other areas of our life, we might "Look to the LORD and His strength; seek His face always" (Psalm 105:4).

I would like to end by encouraging you to behold God in everything you do. Math is only one tiny area in which we can behold God. Every area of our life is, like math, a looking glass through which we can see God's reflection. The principles we have explored in math are equally true in science, history, English, and everything else we learn and do. God's truth is waiting for us *everywhere*. He longs for us to depend on Him in *everything*. May we fix our eyes on Him each day.

But we all, with unveiled face, beholding as in a mirror the glory of the Lord, are being transformed into the same image from glory to glory, just as from the Lord, the Spirit.

2 Corinthians 3:18 NASB

Idea Notebook

Listed below is a sample idea notebook containing a few common ways we use math in our daily lives. See chapter 9 to find out how to develop and use your own idea notebook as you teach.

COMMON WAYS WE USE MATH

Addition and Subtraction

- ❖ Balancing a checkbook
- ❖ Reconciling a bank statement
- ❖ Calculating age (how long a company has been in business, how long the luncheon meat has been in the refrigerator, how old someone is)
- ❖ Dealing with time problems (how long has dinner been in the oven, how long Johnny slept during his nap)
- ❖ Shopping

Algebra

- ❖ Finding any sort of unknown value (Any time we find an unknown value using addition, subtraction, multiplication, or division we are really using algebra. For example, when we try to find the unit price of something, we are trying to find an unknown value. We could use a p, x, or any other letter to represent the unit price. Algebra simply uses a letter to represent the unknown in an equation.)
- ❖ Solving a problem with an unknown number (Sometimes we are

not sure how to find the answer we need. For example, we might be trying to calculate how many bags of mulch we need for the garden. We know that each bag covers 200 square feet and that we need to cover 1500 square feet. Thus we are trying to find what number times 200 will cover 1500 feet or $w \cdot 200 = 1500$. We take 1500 and divide it by 200 or $w = 1500/200$. Algebra is simply a different way of writing equations we do all the time. While most of us do not bother to write down the algebra behind everyday equations, algebra comes in handy when we do not know how to find the answer or when we are dealing with more complex equations.)

❖ Using a formula (finding the area of any shape, calculating interest rates)

Basic Number Functions
❖ Telling time
❖ Reading numbers (radio dials, TV remotes, calculators, baking temperature gauges, thermometers, timers)
❖ Dialing telephone numbers

Decimals
❖ Dealing with money in any form
❖ Paying bills
❖ Balancing a checkbook
❖ Finding prices while shopping
❖ Dividing uneven amounts of any kind

See division for examples of things commonly divided.

Division
❖ Dividing things evenly among family members (such as a bag of candy)
❖ Finding the price per unit (determining which size of spaghetti sauce is cheaper or determining the cost of each can of soda)
❖ Planning (calculating how many pages in a book need to be read each day to finish it on time, finding how much yarn is needed for a

crocheting or knitting project, deciding how much to lay aside each month for a yearly bill)

Fractions

- ❖ Halving or doubling a recipe
- ❖ Cutting a pie or cake
- ❖ Converting between types of measurements (converting teaspoons to tablespoons or using a ¼ measuring cup instead of a ½ measuring cup to measure ½ cup of flour)
- ❖ Mixing solutions appropriately
- ❖ Measuring detergent or medicines
- ❖ Folding paper into halves or quarters for a card
- ❖ Determining yardage necessary for a sewing project

Geometry

See measurement and graphing.

Graphing

- ❖ Reading a map
- ❖ Comparing any sort of results or data
- ❖ Planning a garden or other project

Measurement

- ❖ Home Improvements—
  - ◆ Purchasing new flooring
  - ◆ Computing the area of a floor
  - ◆ Comparing prices between products
  - ◆ Finding the best way to lay out the material to avoid waste
  - ◆ Converting square feet to square yards or vise versa
- ❖ Measuring windows for new drapes or blinds
- ❖ Preparing for new countertops or cabinets (how deep should the cabinets be, will the plumbing fit, what size sink is needed)
- ❖ Installing a light fixture (what size fixture will fit, is the fixture centered on the wall or ceiling)
- ❖ Finding the surface area of the walls in a room before painting it (to

know how many gallons of paint to buy)

- ❖ Landscaping
  - ◆ Measuring the area to be landscaped
  - ◆ Purchasing the appropriate quantity of bark or rocks
  - ◆ Planning for plant growth and deciding how far apart to place plants
- ❖ Around the House—
  - ◆ Hanging a picture (especially if it needs to be centered)
  - ◆ Arranging furniture (knowing if something will fit before it is moved)
  - ◆ Cooking (uses many different measurements and conversions, such as teaspoons, tablespoons, cups, liters, and ounces)
  - ◆ Mixing solutions such as wallpaper stripper or Miracle-Grow (can also involve multiplication if larger quantities are needed)

Multiplication

- ❖ Determining the quantity of items necessary for a project (determining how many pencils are needed in order to provide five pencils per guest for twenty guests)
- ❖ Finding the total spent on something per year (finding how much is spent each year if $18 is spent on trash pick up each month)
- ❖ Shopping

Percents

- ❖ Determining sales tax on an item
- ❖ Computing sale prices
- ❖ Comparing interest rates
- ❖ Grading papers (taking the total number of problems and figuring out the percentage of correct answers)

Rounding and Estimating

- ❖ Shopping (keeping an approximate mental total)
- ❖ Estimating taxes and tax withholdings
- ❖ Budgeting

❖ Managing time (determining what time dinner will be ready if the Crock-Pot started at 8:00 a.m., calculating what time to leave for soccer practice in order to have time to stop at the bank on the way)

Sets and Counting

❖ Organizing (putting like things together and sorting by sets)

❖ Doing the wash (sorting the dirty wash into sets by color or type, sorting the clean wash by family member, matching socks)

❖ Unloading the silverware (dividing into sets of forks, knives, big spoons, and small spoons)

❖ Dealing with money (adding nickels, dimes, or quarters is really counting in sets of 5, 10, or 25)

❖ Counting a variety of objects (library books, people present, number of cookies, time, money, toys)

Common Projects Using Multiple Math Concepts

❖ Sewing (uses fractions, multiplication, division, addition, subtraction, and even algebra)

❖ Woodworking (uses measurements, addition, subtraction, algebra, proportions, and graphs)

❖ Operating a lemonade stand (uses addition, decimals, and division—adding the cost of ingredients uses addition; counting money uses decimals; finding the cost of each glass of lemonade involves division)

❖ Planning a party (uses multiplication and division—multiplication to determine how many items to purchase and division to find the cost per attendee)

❖ Buying a house or car (can involve almost every concept—finding and comparing interest rates involves percents and multiplication as well as decimal numbers; deciding what numbers to divide or multiply uses an algebraic equation)

APPENDIX A

A Few Basic Ways Math Is Used in Science

- ❖ Classifying animals and plants
- ❖ Discovering order in creation (such as the order in sunflowers, honeycombs, red blood cells, and sound waves—see chapter 2)
- ❖ Comparing results of experiments (graph and compare the effects of different lighting on a seed's growth)
- ❖ Predicting reactions
- ❖ Converting measurements (liters to cups or milliliters)
- ❖ Mixing chemicals
- ❖ Designing and engineering products
- ❖ Predicting the results of a proposed design
- ❖ Selecting the best materials to use in a project
- ❖ Recording relationships and natural "laws" (the relationship between pressure and volume in two liquids, the relationship between velocity and distance and time, the Law of Gravity)

Bibliography

MODERN TEXTBOOKS

The reviews given in chapter 10 were based on the textbooks listed below, as well as previews/information on the website listed with the review. Although other math textbooks were consulted, only reviewed textbooks are listed.

Barrett, Everard. *Professor B's Power Mathematics for Children! Workbook One.* 5th ed. rev. Duluth, GA: Professor B Enterprises, 2005.

Beechick, Ruth. *Parent-Teacher Guide for Ray's New Arithmetics.* Milford, MI: Mott Media, 1985.

Boyles, Carlita. *Math on the Level: Pre-K Through Pre-Algebra in One Complete Math Curriculum.* USA: Family Lifestyle Learning, 2007.

Cotter, Joan A. *RightStart™ Mathematics.* Hazelton, ND: Activities for Learning, 2001-2004.

Cummins, Sareta A. *Horizons Mathematics 1 Teacher Handbook.* Chandler, AZ: Alpha Omega Publications, 1991.

Demme, Steven P. *Demme-stration Video.* Drumore, PA: Math-U-See, 1999.

———. Math-U-See: *Foundations of Mathematics.* Drumore, PA: Math-U-See, 1998.

———. Math-U-See: *Intermediate Mathematics Teacher's Manual.* Drumore, PA: Math-U-See, 1999.

Farmer, Letz. *Defeating Division.* Arden, NC: Mastery Publications, 1995.

———. *Finishing Fractions.* Arden, NC: Mastery Publications, 1991.

————. *Mastering Multiplication*. Arden, NC: Mastery Publications, 1990.

————. *Parent's Manual*. Arden, NC: Mastery Publications, 1997.

————. *Subduing Subtraction*. Arden, NC: Mastery Publications, 1990.

Hake, Stephen, and John Saxon. *Math 65: An Incremental Development*. 2nd ed. Norman, OK: Saxon Publishers, 1995.

————. *Math 76: An Incremental Development*. 2nd ed. Norman, OK: Saxon Publishers, 1992.

————. *Math 87: An Incremental Development*. Norman, OK: Saxon Publishers, 1991.

Hannon, Katherine Ann and Adam Floyd. *Principles of Algebra 2: Applied Algebra from a Biblical Worldview*. Green Forest, AR: Master Books, 2021.

Horizons Mathematics K Teacher Handbook. Chandler, AZ: Alpha Omega Publications, 1994.

Howe, Judy England. *Arithmetic 1: Teacher's Edition*. With contributions by Patricia East, Kim Marie Ashbaugh, Cathryn Kauger, and Gloria Rigsby. Pensacola, FL: A Beka Book Publications, 1993.

————. *Arithmetic 3: Teacher Key*. 4th ed. Pensacola, FL: Pensacola Christian College, 1996.

————. *Arithmetic 4*. 1995 ed. Pensacola, FL: A Beka Book Publications, 1995.

————. *Pre-Algebra: Basic Mathematics II; With Problem Solving Strategies, Teacher Key*. 2nd ed. Pensacola, FL: A Beka Book Publications, 1996.

Jacobs, Harold R. *Geometry: Seeing, Doing, Understanding*. 3rd ed. Green Forest, AR: Master Books, 2018.

Jacobs, Tammie D., Susan J. Lehman, Dottie A. Oberholzer, Lynnette Chevalier, Lurene Dubois, Caryn M. Moody, Debra Overly, et al. *Math 2 for Christian Schools: Teacher's Edition*. 2nd ed. Greenville, SC: Bob Jones University Press, 1993.

Jacobs, Tammie D., Susan J. Lehman, Dottie A. Oberholzer, Lynnette Chevalier, Sharon Hambrick, Caryn M. Moody, Debra-Ann Overly, et al. *Math 3 for Christian Schools: Teacher's Edition*. 2nd ed. Greenville, SC: Bob Jones University Press, 1994.

————. *Math 4 for Christian Schools: Teacher's Edition*. 2nd ed. Greenville, SC: Bob Jones University Press, 1996.

Loop, Katherine A. *Principles of Mathematics: Biblical Worldview Curriculum*. Book 1 and 2. Green Forest, AR: Master Books, 2015-2016.

Miller, [Taina] Maria. *Mammoth Math*. NP: Mammoth Math, 2007-2011.

O'Dell, Angela. *Math Lessons for a Living Education*. Level 1. Green Forest, AR: Master Books, 2016.

O'Dell, Angela and Kyrsten Carlson. *Math Lessons for a Living Education*. Levels 2-5. Green Forest, AR: Master Books, 2016.

Pilger, Kathy D., Ron Tagliapietra, Larry L. Hall, and Larry D. Lemons. *Algebra 1 for Christian Schools*. 2nd ed. Greenville, SC: Bob Jones University Press, 1999.

Primary Mathematics Project Team. *Primary Mathematics 1B:Textbook*. U.S. ed. Singapore: Times Media Private, 2003.

―――. *Primary Mathematics 2A: Textbook*. U.S. ed. Singapore: Times Media Private, 2003.

―――. *Primary Mathematics 3B: Workbook, Part 2*. Singapore: Times Media Private, 1999.

Professor B Enterprises. *Professor B Mathematics Power Learning: Online Samples*. Professor B Enterprises. http://www.profb.com (accessed July 2008).

Quine, David. *Making Math Meaningful Level 1 Parent/Teacher Guide*. Rev. ed. Richardson, TX: Cornerstone Curriculum Project, 1997.

―――. *Making Math Meaningful Level 6*. Rev. ed. Richardson, TX: Cornerstone Curriculum Project, 1998.

―――. *My Very Own Making Math Meaningful Level K Parent Guide*. Rev. ed. Richardson, TX: Cornerstone Curriculum Project, 1998.

―――. *Principles from Patterns: Algebra 1*. Rev. ed. Richardson, TX: Cornerstone Curriculum Project, 1997.

Rasmussen, Lore. *Miquon Math*. Emeryville, CA: Key Curriculum Press, 1978.

Rasmussen, Steve. *Key to Fractions*. Emeryville, CA: Key Curriculum Press, 1980.

Rasmussen, Steven, and David Rasmussen. *Key to Percents*. Berkeley, CA: Key Curriculum Press, 1988.

Rasmussen, Steven, and Spreck Rosekrans. *Key to Decimals*. Emeryville, CA: Key Curriculum Press, 1985.

Ray, Joseph. *Ray's New Practical Arithmetic*. Van Antwerp, Bragg & Co., 1877. Reprinted, Milford, MI: Mott Media, 1985.

Sabouri, Greg and Shawn. *Algebra 1: A Teaching Textbook™*. First ed. Oklahoma City, OK: Teaching Textbooks, 2004.

Saxon, John H., Jr. *Algebra ½: An Incremental Development*. 2nd ed. Norman, OK: Saxon Publishers, 1997.

———. *Algebra 1: An Incremental Development*. 2nd ed. Norman, OK: Saxon Publishers, 1990.

———. *Algebra 2: An Incremental Development*. 2nd ed. Norman, OK: Saxon Publishers, 1997.

Schmidt, Stanley F. *Life of Fred: Fractions*. Reno, NV: Polka Dot Publishing, 2008.

———. *Life of Fred: Beginning Algebra*. Reno, NV: Polka Dot Publishing, 2009.

Teaching Textbooks. Online Samples. Oklahoma City, OK: Teaching Textbooks. http://www.teachingtextbooks.com (accessed August 2010).

VideoTextInteractive. *Video and Print Sampler*. NP: VideoTextInteractive, 2009.

OLDER TEXTBOOKS

These textbooks dating from 1768-1930 were the basis for the generalities drawn in chapter 6.

Adams, Daniel. *Adams's New Arithmetic, in which the Principles of Operating by Numbers Are Analytically Explained.* . . . Keene, NH: J & J.W. Prentiss, 1830.

———. *The Scholars Arithmetic; or, Federal Accountant. Containing . . . The Whole in A Form and Method Altogether New, for the Ease of the Master and the Greater Progress of the Scholar.* 5th ed. Leominster, MA: Adams & Wilder, 1804.

Austin, F. E. *Preliminary Mathematics*. Hanover, NH: 1917.

Barker, Eugene Henry. *Applied Mathematics for Junior High Schools and High Schools*. Boston: Allyn and Bacon, 1920.

Bonnycastle, John. *The Scholar's Guide to Arithmetic; or, A Complete*

Exercise Book for the Use of Schools with Notes, Containing the Reason of Every Rule. . . . 9th ed. New York: Isaac Riley, 1815.

Breslich, Ernst R. *First-Year Mathematics for Secondary Schools*. 4th ed. Chicago: University of Chicago Press, 1915.

Brothers of the Christian Schools. *Primary Arithmetic Mental and Written*. New York: La Salle Bureau, 1900.

Fish, Daniel W. ed. *The Rudiments of Written Arithmetic: Containing Slate and Black-Board Exercises for Beginners, and Designed for Graded Schools*. New York: American Book Co., 1905.

Love, John. *Geodaesia; or, The Art of Surveying and Measuring Land Made Easy*. . . . *Corrected and improved by Samuel Clark. London: 1768*. A facsimile reprint of the 8th ed. Schenectady, NY: United States Historical Research Service, 1997.

Newton, Isaac. *Universal Arithmetick; or, A Treatise of Arithmetical Composition and Resolution*. Translated by Mr. Ralphson and revised and corrected by Mr. Cunn. London: W. Johnston, 1769.

Rich, A. W. *The New Practical Arithmetic: Presenting Definitions, Principles, Drills, Model Solutions and Test Problems*. Chicago: A. Flanacan Co., 1900.

Robinson, James. *The American Arithmetic: In which the Principles of Numbers Are Explained and Illustrated by a Great Variety of Practical Questions*. Boston: John P. Jewett & Co., 1850.

Root, Erastus. *An Introduction to Arithmetic: For the Use of Common Schools. Revised, corrected, and enlarged*. Norwhich: Russell Hubbard, 1811.

Seaver, Edwin P[liny]. *Logarithmic and Trigonometric Tables*. 1838.

Seaver, Edwin P[liny]., and George A. Walton. *The Franklin Elementary Algebra*. Boston: William War & Co., 1882.

Stone, John C., and James F. Millis. *A Secondary Arithmetic: Commercial and Industrial for High, Industrial, Commercial, Normal Schools, and Academies*. Boston: Benj. H. Sanborn & Co., 1908.

Yeingst, Wilbur M. *Practical Problems in Arithmetic: Book 1 for Third Grade*. St. Louis: Webster Publishing Co., 1930.

Young, J.R. *The Elements of Mechanics, Comprehending Statics and Dynamics. With A Copious Collection of Mechanical Problems. Intended for the*

Use of Mathematical Students in Schools and Universities. With Numerous Plates. Revised and corrected by John D. Williams. Philadelphia: Hogan and Thompson, 1839.

SUPPLEMENTAL MATH BOOKS

Below are some supplemental math books. See chapter 10 for details about a few of these. Check your local library to see the collection they carry.

Adler, Irving. *The Giant Golden Book of Mathematics: Exploring the World of Numbers and Space.* New York: Golden Press, 1960.

Challoner, Jack. *The Science Book of Numbers.* U.S. ed. Orlando: Harcourt Brace Javanovich, 1992.

Clemson, Wendy, David Clemson, and others. *Using Math to... Series.* Milwaukee, WI: Gareth Stevens Publishing, 2005.

Consortium for Mathematics and Its Applications. *For All Practical Purposes: Mathematical Literacy in Today's World.* Video series currently available on http://video.google.com. Produced in 1988.

Dalton, Leroy C. *Algebra in the Real World.* Palo Alto, CA: Dale Seymour Publications, 1983.

Easterday, Kenneth E., Loren L. Henry, and F. Morgan Simpson, comps. *Activities for Junior High School and Middle School Mathematics: Readings from the Arithmetic Teacher and the Mathematics Teacher.* Reston, VA: The National Council of Teachers of Mathematics, 1981.

Gardner, Robert. *Science Projects About Math.* Berkely Heights, NJ: Enslow Publishers, 1999.

Huff, Darrell. *How to Lie with Statistics.* New York: W. W. Norton, 1993.

Kaye, Peggy. *Games for Math: Playful Ways to Help Your Child.* New York: Pantheon Books, 1987.

Kogelman, Stanley, and Barbara R. Heller. *The Only Math Book You'll Ever Need.* New York: Facts on File Publications, 1994.

Littlefield, Cindy A. *Real-World Math for Hands-On Fun!* Charlotte, VT: Williamson Publishing, 2001.

Markle, Sandra. *Discovering Graph Secrets.* New York: Atheneum Books for Young Readers, 1997.

Polonsky, Lydia, Dorothy Freedman, Susan Lesher, and Kate Morrison. *Math for the Very Young: A Handbook of Activities for Parents and Teachers.* New York: John Wiley & Sons, 1995.

Reimer, Luetta, and Wilbert Reimer. *Mathematicians Are People, Too: Stories from the Lives of Great Mathematicians.* Palo Alto, CA: Dale Seymour Publications, 1990.

————. *Mathematicians Are People, Too: Stories from the Lives of Great Mathematicians. Vol. 2.* U.S.A.: Pearson Education, Inc./Dale Seymour Publications, 1995.

Riley, James, Marge Eberts, and Peggy Gisler. *Stand Up Math: 180 Fun and Challenging Problems for Kids!* Glenview, IL: Good Year Books, 1995.

VanCleave, Janice. *Janice VanCleave's Math for Every Kid: Easy Activities that Make Learning Fun.* New York: John Wiley & Sons, 1991.

Vergara, William C. *Mathematics in Everyday Things.* New York: Harper and Brothers, 1959.

Miscellaneous

A handful of resources on everything from inventions to a biblical view of math.

Aharoni, Ron. *Arithmetic for Parents: A Book for Grownups About Children's Mathematics.* Translated from Hebrew by Danna Reisner. El Cerrito, CA: Sumizdat, 2007.

Cajori, Florian. *A History of Mathematical Notations: Two Volumes Bound as One.* Mineola, NY: Dover Publications, 1993. First published in 1928 and 1929 by Open Court Publishing. Citation is to this Dover Two Volumes Bound as One edition.

Duffy, Cathy. *Christian Home Educators' Curriculum Manual: Elementary Grades.* Westminster, CA: Grove Publishing, 2000.

Educational Research Library (National Institute of Education). *Fifteenth to eighteenth century rare books on education: a catalog of the titles held by the Educational Research Library.* Washington: National Institute of Education: For sale by the Supt. Of Docs., U.S. Govt. Print. Off., 1976.

Gillen, Allen L. *Body by Design.* Green Forest, AZ: Master Books, 2001.

Groza, Vivian Shaw. *A Survey of Mathematics: Elementary Concepts and*

Their Historical Development. New York: Holt, Rinehart, and Winston, 1968.

Hawking, Stephen, ed. *On the Shoulders of Giants: The Great Works of Physics and Astronomy.* Philadelphia: Running Press Book Publishers, 2002.

Hoskin, Michael, ed. *Cambridge Illustrated History: Astronomy.* Cambridge, UK: Cambridge University Press, 1997.

James, Peter, and Nick Thorpe. *Ancient Inventions.* New York: Ballantine Books, 1994.

Knott, Ron. *"Fibonacci Numbers and Nature."* University of Surrey. http://www.mcs.surrey.ac.uk/Personal/R.Knott/Fibonacci/fibnat.html.

Loop, Katherine A. *Principles of Mathematics: Biblical Worldview Curriculum.* Book 1 and 2. Green Forest, AR: Master Books, 2015-2016.

Loop, Katherine A. *Revealing Arithmetic: Math Concepts from a Biblical Worldview.* Fairfax, VA: Christian Perspective, 2009.

Mayall, R. Newton, and Margaret W. *Mayall. Sundials: Their Construction and Use.* 3rd ed. Cambridge, MA: Sky Publishing Corp., 1994.

Nickel, James D. *Mathematics: Is God Silent?* Rev. ed. Vallecito, CA: Ross House Books, 2001.

Reid, Struan. *Invention and Discovery.* Universal ed. London: Usborne Publishing Ltd., 1986.

Schlager, Neil, ed. *How Products Are Made: An Illustrated Guide to Product Manufacturing. Vol. 1.* Detroit: Gale Research Inc., 1994.

The Shorter Catechism. Agreed upon by the Reverend Assembly of Divines at Westminster. *Quoted in The New England Primer, Improved. . . .* Boston: printed by Edward Draper, Newbury Street and sold by John Boyle, Marlborough Street, 1777. Reprint, Aledo, TX: WallBuilder Press, 2003.

Stout, Kathryn. *Maximum Math.* Rev. ed. Wilmington, DE: Design-A Study, 1999.

Zimmerman, Larry. *Truth and the Transcendent: The Origin, Nature, and Purpose of Mathematics.* KY: Answers in Genesis, 2000.

ELEMENTARY SCHOOL SUPPLEMENTAL CURRICULUM

Looking for more detailed examples, ideas, and information to help you teach math from a biblical worldview?

Revealing Arithmetic (also by Katherine [Loop] Hannon) picks up where *Beyond Numbers* leaves off, walking through arithmetic concepts—from counting through exponents and roots. You'll find lots of historical information and practical ideas, as well as specific teaching suggestions, examples of how to change your curriculum, and more!

I refer to it often to keep us from getting bogged down in the daily grind of arithmetical calculations, formulas, rules, "curriculum," etc. *Deidra R., homeschool mom*

Revealing Arithmetic is a spectacular book! It will be required reading for all of my "Math Ed." students. *Jay Auxt, engineer, author, speaker, and adjunct instructor at Antietam Bible College*

I have been studying and thinking about how to acknowledge God in math and science for many years, and this is the first book I have seen that is easy to read with very useful and practical assignments for the younger students, pre-K to about grade 6.
James Bartlett, executive director of the North Dakota Home School Association and Biblical Concourse of Home Universities

Learn more at **www.christianperspective.net**.

Wish there was a curriculum that taught from the perspective explained in *Beyond Numbers*? Well, there is!

Junior High—*Principles of Mathematics: Math Concepts from a Biblical Worldview* (also by Katherine [Loop] Hannon) integrates science, history, real-life examples as it firms up the foundational concepts and prepares students for upper-level math. Students will finish seeing math as a God-given tool that points us to Him and that we can use to serve Him wherever He calls us. Perfect for grades 6-8, or as an overview course for older students. There's even an optional eCourse to go alongside the program where Katherine teaches the material.

Algebra 2—Katherine and her husband, Adam, wrote a complete program, *Principles of Algebra 2: Applied Algebra from a Biblical Worldview*, that actually shows students why they're learning algebra...and how algebra proclaims God's praises. There's an optional eCourse component featuring a video for each lesson where Katherine teaches the material.

...OR KATHERINE'S OTHER ECOURSES!

In addition to eCourses to go with the *Principles* curriculum, Katherine has also produced the following eCourses.

Jacob's Elementary Algebra (Algebra 1)—Algebra can feel so over-whelming...but this course brings it to life! Katherine teaches each lesson from a biblical worldview. Students will leave awed at the Creator and equipped with what they need to succeed at algebra.

Jacob's Geometry: Seeing, Doing, Understanding—See geometry like you've never seen it before! The videos brings the textbook to life, walking students through the lessons and giving tips on problems and how to master the material. Above all, the eCourse integrates a biblical worldview throughout the program, leaving students encouraged and challenged spiritually as they study.

Revealing Arithmetic— These videos serve like a travel agent, showing different ideas you can incorporate into your math lessons, including fun and hard-to-find historical information.

Learn more at **www.christianperspective.net.**

ABOUT THE AUTHOR

Katherine (Loop) Hannon has been writing and speaking about math for more than 15 years. Understanding the biblical worldview in math during her senior year of homeschool made a tremendous difference in her life and started her on a journey of researching and sharing on the topic. Her books on math and a biblical worldview have been used by various Christian colleges, homeschool groups, and individuals.

Katherine lives in Georgia with her husband, Adam. She is active in facilitating women's Bible studies and reaching out with the gospel. She enjoys music, reading, spending time with friends and family, and learning/exploring new things. Her favorite thing to do is to talk with people about God and His Word. After all, all our biographies will end with seeing Him face to face.

❖ **Website and Biblical Math Blog**
   www.ChristianPerspective.net
❖ **Facebook** - www.facebook.com/KatherineLoopHannon
❖ **YouTube** - www.youtube.com/user/MathIsNotNeutral

www.ingramcontent.com/pod-product-compliance
Lightning Source LLC
Chambersburg PA
CBHW060618210326
41520CB00010B/1380